"Sarah DiGregorio delves d͟e͟e͟p͟ ͟i͟n͟t͟o͟ ͟t͟h͟e fraught world of premature birth. With bracin͟g͟ ͟.͟.͟.͟.͟.͟.͟.͟.͟.͟.͟.͟.͟.͟.͟.͟.͟.͟ tory and the stories of other w͟.͟ love and meld it with cutting-e͟.͟.͟.͟.͟.͟.͟.͟.͟.͟.͟.͟.͟.͟.͟.͟.͟ e the life of their newborn. Thi͟.͟.͟.͟.͟.͟.͟.͟.͟.͟.͟.͟.͟.͟.͟.͟.͟s to a world that is rarely seen with such clarity."

> —Jerome Groopman, MD, Recanati Professor of Medicine,
> Harvard Medical School, author of *The Anatomy of Hope*

"A must-read for anyone interested in the science—or the experience—of preterm birth."

> —Emily Oster, author of the *New York Times* bestseller
> *Cribsheet* and *Expecting Better*

"The heart of DiGregorio's illuminating book isn't just about her family's journey; it's an expansive examination of the history and ethics of neonatology."   —*New York Times Book Review*

"Fascinating. DiGregorio has strung together a riveting history of the preemie, from carnival incubator shows (really!) to the possible future of baby ziplocks. Throughout, she has tenderly woven her personal experience with her tiny daughter in the NICU, a space where machine and mother uneasily coparent. At times shocking, heartbreaking, and inspiring, the tension between technology and humanity is evident throughout, and DiGregorio does not shy away from it."

> —Jennifer Block, author of *Everything Below the Waist*

"A sweeping cultural history, a consistently surprising and insightful examination on the porous line between life and death, and a graceful and hauntingly clear-eyed memoir all in one. Feels destined to live on shelves for a long time."

> —Jayson Greene, author of *Once More We Saw Stars*

"Every health professional who cares for babies should read this book. A meticulously researched and stirring tribute to the life-affirming work that goes on in NICUs every day."

—Dr. John D. Lantos, pediatric bioethicist and author of
Neonatal Bioethics: The Moral Challenges of Medical Innovation

"Sarah DiGregorio's rigorous, gimlet-eyed reporting on premature birth is surpassed only by her empathy and affection for the people whose lives comprise its history and give it meaning. How can we better care for the most vulnerable people in our communities who, as DiGregorio makes clear, include not only the tiniest babies but also the adults who come before them? . . . She triumphs at making issues typically confined to the NICU relevant to every human being." —Angela Garbes, author of *Like a Mother*

"DiGregorio makes clear that the problems facing preterm babies can be enormous, that consequences may not be apparent for years, and that the appropriateness of treatment can be debated. . . . Clear reporting that wisely urges careful decision-making by clinicians and parents alike." —*Kirkus Reviews*

"Compassionate. . . . Sensitively approaching the myriad practical and ethical challenges involved in caring for such fragile babies, DiGregorio gives vivid, individualized portraits of struggling parents, premature infants who developed into thriving children, and the specialists dedicated to helping them. . . . DiGregorio delivers a candid yet gentle work with appeal for prospective parents and anyone interested in 'what premature birth [can] teach us about being human.'" —*Publishers Weekly*

Early

Early

An Intimate History of Premature Birth and What It Teaches Us About Being Human

SARAH DIGREGORIO

HARPER

NEW YORK . LONDON . TORONTO . SYDNEY

HARPER

HarperCollins books may be purchased for educational, business, or sales promotional use. For information, please email the Special Markets Department at SPsales@harpercollins.com.

Designed by Bonni Leon-Berman

Epigraph on page 305 from *The Ground Beneath Her Feet* by Salman Rushdie, copyright © 1999 by Salman Rushdie. Reprinted by permission of Henry Holt and Company in the United States. Reprinted by permission of Vintage Canada/ Alfred A. Knopf Canada, a division of Penguin Random House Canada Limited. All rights reserved. Any third party use of this material, outside of this publication, is prohibited. Interested parties must apply directly to Penguin Random House Canada Limited for permissions.

Library of Congress Cataloging-in-Publication Data has been applied for.

ISBN 978-0-06-282031-0 (pbk.)

22 23 24 25 26 LSC 10 9 8 7 6 5 4 3 2 1

For Phyllis and Mira, my mother and my daughter,
who taught me how to be brave.

And with gratitude to the nurses, physicians,
and thinkers who were our partners in gestation.

Contents

Author's Note   ix

Prologue: One Birth   1

Part I: The Unexpected: Millions of Births

1. What Happened?   31
2. Treatments and Outcomes   39
3. Viability and the Zone of Parental Discretion   45

Part II: The Body: Incubation

4. The History of Incubation: Coney Island,
   Chicken Eggs, and Changelings   51
5. The Modern Incubator, or How to Build a Giraffe   74
6. The Incubators of the Future: Babies in Bags   83

Part III: The Breath: Treating Respiratory Distress

7. Dr. Mildred Stahlman and the Miniature Iron Lung   99
8. Dr. Maria Delivoria-Papadopoulos and the Rugged Machine   112
9. JFK's Lost Baby and the Advent of Surfactant   121

Part IV: The Self: Protecting the Premature Brain

10. The Revolutionary Practice of Listening to Preemies   133
11. Follow-up Care: Preemie Development Beyond the NICU   153

Part V: The Threshold: End-of-Life Issues at Birth

12. What Should We Do for 22-Week Babies?   177
13. Knowing When to Stop   203
14. Choice, Decisions, and the Messiness of Real Life   227

Part VI: The Crisis: The Body Under Stress

15. Racism Causes Preterm Birth   243
16. What Prematurity Means in Mississippi   260
17. Group Prenatal Care and the Power of Community   271

Part VII: The Invisibles: Breaking the Silence

18. The Hidden Trauma of Prematurity   289
19. Grown Preemies Speak for Themselves   297
    Epilogue   306

Acknowledgments   *309*
Notes   *313*
Index   *337*

Author's Note

Throughout this book, for flow and simplicity, I refer to preterm babies as preterm, premature, or, more casually, preemies. The current medical term is "preterm," not "premature," but, since they are both commonly used in the vernacular, I use them both.

In some cases I refer to a fetus as a baby, because that is how some people think of their fetuses, especially as the pregnancy progresses. In some instances, I have used it to accurately describe the way parents thought of their pregnancies.

In places, there are references to "pregnant women" instead of "pregnant people," which might be read as conflating womanhood and the biological capacity for pregnancy, which is not always the case. Trans men and nonbinary people can also get pregnant and experience everything that pregnancy might entail, and I hope people of all genders will feel included by this book.

If there's one thing I have learned, it's that good physicians do not always agree; in fact, they more often disagree. The scientific and medical information in this book is as accurate as possible at the time of this writing—it changes all the time—but it is not comprehensive. I have not covered every possible iteration of preterm birth, nor is this a book to turn to for medical advice. If you are in need of medical advice, the very best thing you can do is ask your clinician, who knows your child, who is unique in the world. Show up with a notebook and a pen, and don't be afraid to keep asking questions until you understand.

One name has been changed, and several last names omitted, to protect privacy. Some details have been omitted for the same reason. But nothing inaccurate has been added. Interviews have been condensed and edited for clarity.

Finally, there are parts of this book that might not be comfortable for pregnant people to read. There are parts that might be hard on parents whose babies are still in the neonatal intensive care unit (NICU). It's true that giving birth prematurely is frightening and there can be long-term health implications. But the larger, overarching truth is this: The vast majority of babies born prematurely go on to live happy, healthy lives. No one wants to see their child face challenges. But the better information and support you have, the easier everything will be.

Early

Prologue:
One Birth

The NICU [is] a strong, strange, powerful place.

There is probably no eerier place in a hospital than the NICU. One enters thinking that one is prepared to see tiny babies. But the babies are unimaginably tiny. They are magical. . . . The babies seem almost, but not quite, human, almost, but not quite, fetal. In their chimerical, half-human, half-machine state they seem not only helpless and pitiful but also exotic, threatening, futuristic, feral, untamed, barbarous . . .

[And yet these] are just babies. . . . They are real little people, and this spaceship, this high-tech roller coaster, this cyberwomb, is their introduction to life on earth.

—DR. JOHN D. LANTOS, *THE LAZARUS CASE: LIFE-AND-DEATH ISSUES IN NEONATAL INTENSIVE CARE*

THIS ISN'T THE story of one birth; it's the story of millions of births. But, for me, it started with Mira.

The first time I saw my daughter she was wrapped in a pink-and-blue-striped hospital blanket, the same one that bundles every infant in every photo on Facebook. She was briefly held aloft for me to see by a nurse who was hustling her out of the operating room.

My memory of the moment is unreliable. I know she was intubated, but I don't remember the tube down her throat, as though some kind of censor in my mind has blacked it out. I can't remember anything about the way she looked, only the impossible, science-fiction smallness of her, a 1-pound, 13-ounce baby, the blur of scrubs in motion, and the way my husband's face crumpled like a piece of paper at the sight of her, the way his head fell into his hands. Separated from my numbed bottom half by a hanging blue sheet, I felt weirdly calm, set apart from the proceedings. I had only one thought, like a beat: *She's alive. She's alive. She's still alive.*

My body had been trying to kill her. Months earlier we were sitting in a young radiologist's office when she pushed a printout of blood work results across the lacquered desk to us. "There's a small chance, maybe a 1 percent chance, that she'll be a little early, a little small," said the doctor, pointing to one line on the sheet. She had just completed our twelve-week ultrasound and blood work, and everything was fine except for one abnormality: very low PAPP-A— the jaunty nickname for pregnancy-associated plasma protein A. The lack of it suggested a subpar placenta, the organ responsible for delivering nutrients and oxygen to a baby in the womb. My (our?) level of this protein was in the second percentile. "After twenty-four weeks, we'll have you come in for growth scans to make sure she stays on her growth curve," said the radiologist. "But it's nothing to worry about." *Nothing to worry about; nothing to worry about.* The refrain of my pregnancy.

Of course, I went back to my office and looked it up. Like the extremely responsible Googler that I am, I skipped the many results for pregnancy message-board threads with subject lines like "Low PAPP-A, how worried should I be?????" and went straight for the scientific literature, all of which went something like this: ". . . associated with stillbirth, infant death, intrauterine growth restriction, preterm birth, and preeclampsia." It was all associations, not causations: nothing certain, but nothing good.

I went outside to escape my cubicle. I stood in the middle of the sidewalk, buffeted by bodies. In midtown Manhattan, there's nowhere good to cry, so I balanced on a fire hydrant and sobbed. Two months earlier I had had a miscarriage at ten weeks and a painful dilation-and-extraction procedure to remove the dead fetus from my uterus, which seemed inclined to not give up. I sat up from the stirrups, noticed that my socks were splattered with a horror movie's worth of blood, and started to understand in a bone-deep way that having a child was not going to be easy. I couldn't imagine losing another one. But then, no one can.

At my next obstetrician appointment, my preferred OB, Dr. M., was away. The doctor covering for her wanted to talk about our risk for Down syndrome. It was odd, because although that blood serum test, the one that indicated low PAPP-A, can be used to assess a risk of a chromosomal abnormalities, I had also had a fetal DNA test, which identifies conditions like Down syndrome with better accuracy. The DNA had come back normal—"Normal female," to be exact. (I was surprised how blissed out I was at the news that the baby was a girl.) Nevertheless, the doctor went over the fact that low PAPP-A is associated with chromosomal problems, even though, as far as I could tell, we knew that outcome wasn't at all likely. I opted not to have an amniocentesis, which would have definitively ruled out a chromosomal syndrome, but which also carries a small risk of miscarriage. As for the other worries—the intrauterine growth restriction (IUGR), the preterm birth, stillbirth—I wasn't able to extract any meaningful information. "I've had lots of patients with low PAPP-A who went on to have healthy, full-term babies," the doctor said. "We're going to monitor growth after twenty-four weeks. Try not to worry."

SHE MEANT IT. It was only one ominous test result, and there was no other indication that anything was amiss. But what I didn't know then was that when doctors kept saying, "Don't worry," they really meant two things: One, there probably really was nothing to worry about. Two, there was no point in worrying, because there was no treatment, no preventative, for what might happen. For what was about to happen. Not worrying was the best we could do. But I couldn't manage it.

The weeks ticked by slowly: 14, 15, 16, 17, 18. Those stupid apps: Your baby is the size of a nectarine, an avocado, a pear, a sweet potato, a mango. At least, you hope she is. I tried not to worry. From about week 10 on, I had migraines that lasted days—or, rather, it seemed to be one long migraine that occasionally let up for a couple of hours. I wore sunglasses at work to look at my computer and strapped ice packs around my forehead. I dreamed of the day that I'd be able to take fistfuls of Advil again. But I also loved being pregnant. I loved the slippery flutters when I began to feel Mira move. I loved thinking about having a girl, the idea that I'd get to have another mother-daughter relationship, since my own mother died when I was in college. I thought about the books we'd read together. And I loved eating for two.

At the time, I was a food editor at *Food & Wine* magazine. My favorite selfish pleasure took on a comfortably virtuous cast. It was all for the baby! I drank blueberry-kale smoothies for breakfast and snacked on fat salted cashews. At work I tasted everything the test kitchen churned out: lamb tagine and coconut cake, bitter escarole salad and spiced black-eyed peas. Buttery rolls, homemade croissants. Fried oysters, roasted broccoli, and seafood gumbo. I thought of everything I ate as an experience she and I were having together. I imagined my fetus fat and happy, smacking her tiny lips, perfectly nourished, and destined to be an adventurous eater.

My husband, Amol, and I met in college. We are both only children who were raised in small-town Massachusetts. On our first date we talked for hours about the Red Sox and our early childhood memories of the 1986 World Series heartbreaker. We laughed a lot. It seemed like we were best friends the moment we met. He sustained me through my mother's illness and early death and my father's protracted struggle with both mental and physical illnesses.

We had been undecided on having a baby until we weren't. We always had so much fun together, and suddenly it seemed so obvious that we should say *yes*. And when I was pregnant the first time, before the miscarriage, we were both giddy, and surprised to be giddy. I knew quite well that more love inevitably meant more vulnerability and sadness, too. But I wanted to be brave enough for all of it—brave enough to be someone's mother.

At 20 weeks Mira was in the 23rd percentile for size and everything looked fine. At 24 weeks she had fallen into the 13th percentile. It wasn't good news—falling off her growth curve could presage a problem with the placenta, which was the thing we were dreading—but it wasn't catastrophic, either. So we waited and tried not to worry.

"I think she's had a growth spurt," I said to Amol on our way to our 28-week scan. It was uncharacteristically optimistic of me, and also incorrect. As soon as the wand hit my belly, I could tell by the ultrasound tech's face that the news was not good. It was November 26 and I was due February 16. The tech disappeared to talk to the doctor.

"She's not growing," I said.

"Don't jump to conclusions," Amol said. "We don't know any-thing yet." I did, though.

We were ushered into another office, another lacquered desk, another doctor. Mira had indeed stopped growing entirely. She was below the first percentile; she was off the charts. She was approxi-mately the size of a 26-week fetus. She was going to have to come early. How early was not clear. We had to go to the hospital. I had to get a steroid shot. I could not go back to work. ("What do you *mean* I can't go back to work?" I heard myself say. "I *have* to go back to work.") I could not so much as go for a walk.

We caught a cab to the NYU Langone medical center. On the way I called my cousin, who is a high-risk obstetrician. "How many weeks?" she asked. Twenty-eight and one day. "How many grams are they estimating?" Eight hundred. There was a long pause. "Okay," she said. "Okay. You need the steroid shot."

We had to go up to labor and delivery. We stood in silence, wait-ing for the elevator, until I realized I was no longer supposed to be standing whenever I could help it. I sat on a bench. "Nothing's going to be normal now," I said.

"Here we go, chicken," said a nurse, jabbing an industrial-sized needle into my thigh. That was steroid shot number one. A fetal heart monitor on a thick belt was strapped around my less-than-impressive belly. Mira's heart rate was decelerating, speeding up again, and then decelerating again, so they admitted me. I signed a consent for an emergency C-section, but it wasn't clear when it would happen. A nurse came in and said they needed to start a magnesium drip, which would help protect Mira's brain if she came soon. Amol ran out for a sandwich and the nurse hooked up the bag to my IV. Seeing that my husband was gone, she made some excuse to stay in my room and do paperwork while the magnesium started up. Suddenly I felt like I was burning up from the inside out. I couldn't catch my breath. "I think I'm allergic to this," I said to the nurse. She disappeared and came back with a resident, who checked me. I was fine. "It's just the mag, honey," said the nurse. "It makes you feel like that." It was a twelve-hour course; I sweated and tossed through the night.

So there I was, sweaty and confused, when, early the next morning, a neonatology fellow dispatched from the NICU came in to tell us what to expect when you're expecting a premature baby. At that point no one knew if I would need to deliver in hours, days, or weeks, but there seemed to be no hope of getting to term. Mira was diagnosed with severe intrauterine growth restriction as a result of "unexplained placental insufficiency," a fancy way of saying that no one knew why the placenta was shutting down. I was not providing her with enough nutrition to grow. Soon I might also deprive her of oxygen. I was a well-fed food editor, and my fetus was starving.

The doctor perched awkwardly by the bed and vomited up a litany of potential complications that arise from being born too soon. Bleeding in the brain, holes in the heart, butterfly-wing lungs that struggled to inflate, intestines that died while the baby still lived, blindness, loss of IQ, attention issues, disabilities of all sorts, infection, cerebral palsy, death. He said we had roughly a 50-50 chance of getting out of this without a disability of some kind. Did we have any questions?

It was the first real information I had about prematurity, and I felt like I was drowning. I remembered that a pregnancy book I was reading had said that a 28-week baby has a 90 percent chance of survival—so I asked: Was that right? The doctor blinked. "Well, no, because she's so small, more like the size of a 26-weeker, and because she's been so stressed in utero, her odds of survival will be some-where between a 26- and 28-weeker." What did that mean? Eighty-five percent? Eighty-eight? I didn't know, but I didn't ask again.

"But," he went on, brightening, "premature girls tend to do better than boys. And African Americans tend to do better than whites." Amol and I took this in. (The evidence that Black babies tend to fare better than others is actually mixed and not at all conclusive; on the other hand, there is new evidence that Black and Latinx babies are more likely to get inferior NICU care.) "What about half-white, half-Indian girls," I asked. "Does she get a little bump for being bi-racial?" No one laughed.

It was Thanksgiving Day, and I had gotten two steroid shots and the course of magnesium. Mira's heart rate had steadied, so af-ter scans upon scans that showed the umbilical vessels were still working to keep Mira in oxygen, the doctors decided they could discharge me home. The goal was to stay pregnant for as long as possible. My obstetrician, Dr. M., whom I loved, said the goal was 32 weeks. But the goal was also to keep Mira alive, and the two aims were not necessarily compatible. At some point—no one knew exactly when—my placenta, which I imagined as a beat-up old car, chugging along, belching smoke, would simply stop working, and she would suffocate and die. The high-wire act was to keep Mira inside until the last possible moment and then get her out. So they sent me home, but I needed to be on bed rest and I had to count kicks. If I couldn't feel Mira moving, I had to come back to the hos-pital immediately.

For the two days that we were home, I lay on the couch and Mira moved more than she ever had before. She flipped and flopped. I allowed myself to imagine that everything might be okay. And then on Sunday, November 30, she just stopped. Amol was at Ikea, in a frenzy of baby-room buying and assembling. I ate a cookie and drank a glass of juice, because a sugar rush supposedly wakes a sleeping fetus. I moved around and tried to rouse her. "No kicks," I texted Amol. "Fuck," he texted back.

Back we went, through the Brooklyn-Battery Tunnel, up the FDR Drive, to the hospital, a drive that would soon be too familiar. I was weirdly calm, which is not like me. Amol looked nervous, which is not like him. After we parked the car, he was half jogging to get inside the hospital, and I said, "I don't think it's an *emergency*." He gave me a funny look.

Back up to labor and delivery. A nurse behind a desk. "I'm 28 weeks and I'm not in labor," I announced. "So, why are you here?" she asked. "Oh, no fetal movement and IUGR," I said. They put me in a bed behind a curtain. The woman on the other side of the curtain was in made-for-TV labor, panting and moaning.

On ultrasound, Mira's heart was still beating, but that was the only sign of life. She wasn't moving; her tiny hands were slack. Her heart rate was completely steady—ominous, because heart rates are supposed to be variable; it's a sign that the central nervous system is active. It was, I later learned, a category III fetal heart rate tracing, which necessitates delivery. It means death or brain damage is an imminent risk; there is no category IV.

The obstetrics resident asked for a second opinion, and another, older doctor with a formidably serious countenance came in, looked at the heart rate tracing, took control of the gel-slicked wand, and stared at the motionless fetus on the monitor for a moment. There was no choice presented, for which I am grateful, because I could not really grasp what was happening. Being born nearly 12 weeks early is bad, but being stillborn is worse. The doctor turned to me and said, "Things are going to happen very fast now."

I lay back and covered my eyes with my hands. "Now?" I asked. "Right now?" A nurse was already taking my clothes off, putting a gown on me, finding a vein for an IV.

And just like that, I was swiftly wheeled toward an operating room with what seemed like dozens of doctors and nurses trotting along beside. Dr. M. was on call, and she came quickly down the hallway. "What's the presentation?" she asked. "Transverse," someone else said. Amol said later that it was like being in a car crash: the cold clutch of fear, the way time seems to slow in a sickening, unnatural way. A nurse pulled him aside; he'd have to wait outside the operating room while they set up; he needed to put on scrubs.

Someone warned me that they might not have time for an epidural—they were still looking for an anesthesiologist—in which case they'd just put me under. But as we got to the operating room, an anesthesiologist appeared out of nowhere and said he could do it. A nurse let me put my arms around her as he threaded the needle into my spinal column.

I was half crying into an oxygen mask that had been hastily strapped over my face. The oxygen was for Mira's benefit; she was perfectly still deep inside me. I lay back and my bottom half went heavy. Amol was still outside. "If something happens to me, can you tell my husband I love him?" I asked a nurse. "Oh, honey, we're worried about your baby, not you," she replied. I had a twist of shame. I knew that. But it was hard for me to tell the difference between us, to locate the threat of death, our bodies still knotted together.

"Incision!" said Dr. M. A few minutes of tugging later she called out, "It's a girl!"—which felt like a gift, a moment to pretend. The neonatal team snatched her up. Silence.

I found out later that there had been nineteen clinicians there in the operating room, and for a while all I could hear was a low murmur of voices, a shuffling of feet. There was Dr. M., beyond the blue curtain, rummaging around in my empty uterus. There was the blinding flare of the lights above. There was Amol, wide-eyed and mute in his blue scrubs and hair net, sitting next to me. There was the neonatal team, huddled around a warmer. "Is she alive?" I asked the silence. "Yeah, they're working on her," said a nurse. It wasn't quite an answer. Someone pulled the oxygen mask off my face.

Mira's medical record tells me that the cord was wrapped around her torso and neck and the amniotic fluid was stained with meconium. The cord was cut and she was immediately handed to the NICU team and brought to a warmer. "Limp, no spontaneous breaths," reports the record. She was blue. "Stim [stimulated] and dried but no improvement," it goes on. "PPV [positive pressure ventilation] started. Intubated in delivery room."

Once the ventilator was breathing for her, Mira stabilized. "Color and O_2 status slowly improved," says the record. There was a noticeable change in the air in the operating room; she had been successfully resuscitated, the first and most important of a long series of steps that would have to go right. A nurse took a photo of Mira; in it her eyes are squeezed shut and she is grimacing around the tube in her mouth. Lying on the operating table, I couldn't see her, but I said her name out loud—"Mira"—so she would know I was there. And then, as I was being sewn up, she was whisked away.

What strikes me now is how much human ingenuity and skill and technology was marshaled there to save her, and how utterly helpless and, in a way, incidental, she and I both were. My body was unable to sustain her; her body was unable to sustain itself. She was not so much alive as in limbo.

Afterward, I was confused. Everything seemed to be moving

slowly and strangely, reality twisted and distorted. My body felt like an empty house that had been vacated in a rush, leaving dirty dishes in the sink. I couldn't figure out why I was bleeding from between my legs, since I seemed to have had an operation on my stomach. The muscles in my lower abdomen were twitching violently. It seemed suddenly crucial that I delete the pregnancy app on my phone, since I was no longer pregnant, and my baby was definitely not the size of a butternut squash. I remember a doctor plopping my placenta into what I thought was a takeout container for noodle soup. (It was actually a lab container.)

During that first hour, we were not allowed to see Mira. After they stabilized her in the NICU, Amol was able to visit. The photos that a nurse took of that first father-daughter encounter showed Amol bent over a riot of tubes and lines that hid our 1-pound, 13-ounce daughter.

I had to be able to stand up and get into a wheelchair without fainting before I could go to the NICU myself. In the middle of that first night, after a few false starts—one of which ended in a full-on blackout—I managed to plant myself in the wheelchair. I remember what seemed like a cold breeze on my face; the wheelchair seemed to be moving very fast down a very white hallway, into an elevator, down to the ninth floor, then another white hallway, shoes squeaking. As the automatic doors swept open, I had a sudden sense that I wasn't ready, that this could not be happening, and an equally strong and contrary urge to get to her, find her, see her.

Amol pushed me down a hallway, past rows of incubators, to a hushed, darkened back room, where Mira lay on her back inside what looked like a space-age pod, immobilized by the ventilator that rhythmically inflated her skeletal chest. She was naked except for the world's tiniest diaper; her body was reddish, her forehead creased in what looked like discomfort or worry, lots of fine black hair tufting out of the tiny knit cap on her head. Her blunt little nose was so obviously Amol's that we had recognized it on ultrasounds. Her hands looked long and elegant, her feet strangely large next to her emaciated legs. Her still-forming ears were fused to the side of her head in whorls. She didn't seem to have nipples yet; weeks later they just appeared one day. Her torso was covered with sticky sensors that monitored her vital signs; she had an IV line in her umbilical stump. There were more leads, lines, and tubes than baby.

"You can touch her," a nurse said, popping open one of the portholes for me. I could touch her? I put my hand into the warm incubator and gingerly placed my index finger on the sole of her one-and-a-half-inch foot. There is a photo of that moment, me in a hospital gown looking down at her through the plastic. She is only a foot or so away, but I look like I am staring into the far distance.

I didn't think of it at the time, but I had been in a NICU once before.

There is another photo, this one a Polaroid from 1979: It is of my mother and father standing in front of an incubator, this one more glass box than spaceship. My mother, in her own hospital gown, is reaching through the portholes to cup my body in her hands. An IV protrudes from the top of my head. I was 3 pounds, 14 ounces, 2 pounds more than my own daughter would be at birth.

The look on my mother's face in the photo is identical to my own, across thirty-five years: love, terror, and exhaustion, the same cocktail coursing through the veins of most brand-new parents. But something else, too: not guilt, exactly, but something close to it— helplessness. Everyone realizes eventually that they can't protect their children from suffering. A premature birth is a particularly crushing introduction to that concept. Fundamentally, a mother's body is supposed to be able to cradle and sustain a fetus until it's fully built: ready to breathe air, drink milk, be held.

I don't know much about the circumstances of my own birth, and my twenty-something self didn't ask my mother the questions my thirty-something self would have, had she lived. I know I was due in November but born in September, likely around 32 weeks' gestation. I was dangerously anemic and jaundiced, and I needed to be transferred by ambulance to Women & Infants Hospital in Providence, Rhode Island, for an exchange blood transfusion, in which all of the blood in a baby's body is swapped out for donor blood a little at a time. One possible explanation for both the preterm labor and the extreme jaundice—one that I can't verify because both my parents are dead and the medical records are gone—is Rh disease, a condition in which, because of a mismatch in blood type, a mother's immune system attacks a fetus's blood. It is as though the mother is allergic to the baby. (Rh disease is now treatable with medication the mother can take during pregnancy.)

In both my mother's pregnancy and mine, there was nothing wrong with us or our babies. None of us were sick. (In this, of course, we were lucky.) It was the pregnancy itself—the organism of us together—that went wrong. For my mother, my daughter, and me, the only cure was the end of the pregnancy and the clumsy, miraculous gestation science could provide.

The NICU is both futuristic and primal. It's a place where babies the size of your hand are saved by some of the most advanced technology in the world, but also where all the wizardry of twenty-first-century medicine is a crude and ineffective substitute for a human uterus. Sometimes it is a place where parents hold their babies for the first time only when it's been decided to let them die. It's a place where we, the mothers, sit next to the pods that are doing the work our bodies should have done: breathing for, warming, and feeding our babies.

In the soup of postpartum hormones, it hurt me physically to look at Mira. I found myself taking one painful breath and then another, unable to do anything except live from one second to the next.

The first few days of an extremely low-birth-weight baby's life are critical. Mira was in the back of the NICU, where the smallest, sickest babies were sequestered. No one could say why the placenta had failed, but it could mean an infection or a genetic abnormality. She was put on antibiotics, just in case, and held under blue lights to counter her high bilirubin count. She wore a little mask to protect her eyes from the lights. Between that, the ventilator, and the tape holding it all in place, most of her face was obscured. She flailed and jerked and shuddered, her little hands reaching and stretching. For the first few days she got intravenous nutrition through a central line straight into a large vein near her heart before a feeding tube was slipped down her throat and secured in place with more tape on her chin. She had one-third of a cup of blood in her entire body.

Every morning the attending neonatologist, the fellows, the residents, and the nurses gathered around each NICU bedside one by one and summarized each baby's status and the plan for the day; it's called rounding. On December 2, the morning of her third day, the day I was to be discharged, I sat in my hospital gown next to her pod clutching a notebook and wrote down everything Dr. K.—the attending neonatologist, kind, patient, petite, with a ramrod-straight bearing—said about Mira. My notes begin: "RDS: premature lung disease (???). 780 grams. Peeing a lot. Caffeine. (???) BP stable."

Dr. K. explained that almost all babies Mira's size have a lung disease called respiratory distress syndrome, which means simply that her lungs were not mature enough to breathe on their own, lacking surfactant to keep them inflated. They are also given caffeine as a respiratory stimulant—the medical version of a sharp pat on the cheek: *Stay with us!*

A resident looked at me, clutching my pen, writing furiously. "Oh, Mommy's taking notes," he said, and chuckled. I wanted to tell him that I was an editor, or had been.

Later that day I was able to hold her for the first time, a practice called kangaroo care, in which the parent's bare skin against the baby's bare skin helps them stay warm out of the incubator. Our nurse gently extracted Mira from her pod, trailing all her lines, tethers, and tubes behind, and carefully placed her under my hospital gown, on my chest. She was a collapsible, bony, furred warmth; she curled up with her head beneath my chin, her legs between my breasts. My vision wobbled, but not from tears; I had the sensation of being underwater, of being put back together.

Born emaciated—the term in her chart was "fetal malnourishment"—Mira continued to lose weight. Her skin wrinkled and hung off her toothpick bones. When a nurse changed her diaper (more a scrap of plasticky cotton than anything else), I was horrified to see that she had no bum, none at all. Just legs ending in a bony area with a rectum. She seemed in obvious discomfort, painfully exposed. I wanted to unzip my body and stuff her back in. "I'd prefer she not fall below 700 grams," said Dr. K. when Mira weighed in at 720 grams (1.5 pounds) on day 4. "How do we keep her from going below 700 grams?" I asked. There was no answer because there was no answer.

The sight of a one-and-a-half-pound baby short-circuits something in the brain. There's no roundness, no eye contact, no burbling. No baby-ness. Newborns evolved to be sweet and cuddly, a way to get us to take care of them and ensure the survival of the species. Lots of parents say they have upsetting difficulty in bonding with their preterm infant, at least partly because we haven't evolved to connect with babies that look and act like this.

We don't even really have a word for them, these children, these tiniest of people, who are not fetuses but are not quite babies yet. Being born didn't really make Mira into a newborn. I loved her, and she was my child, but she wasn't quite a baby, or at least not like any baby I had ever imagined. Her brain, if I could have seen it, looked like an almost-smooth lima bean, with only a few ridges and gyrations. The furrowed gray matter called the cortex—the mammalian brain region responsible for language, memory, sensory processing, and almost everything we think of as our humanness—was still developing.

I was sure she would die. At our first family meeting with the doctors, I asked it point-blank. *You're right. She* could *die,* they said. But she was doing well, all things considered. She'd quickly graduated from the ventilator to the CPAP (continuous positive airway pressure), a kind of oxygen mask like one the world's tiniest fighter pilot might wear. It delivered pressurized oxygen to her nose, helping her breathe. On day 5, she finally stopped losing weight and started to gain it, gram by gram. A tiny bit of breast milk—one, then two, then three milliliters—was going down her tube and being successfully digested. Her first brain ultrasound had come back clean, no bleeds, knock wood. Did I want to talk about my fears about disability, the future? I did not. All my conceptions about the future had evaporated.

I imagined making deals with a witch. You might be able to have your heart's desire—your living child—but there will be a cost, now or later. If you're lucky enough to have snatched her back from death, someday there will be a knock on the door, and it will be the witch, hand outstretched for payment.

For most of us, the lucky ones, it's a no-brainer, this deal. *You can save my baby but she may have asthma later? A motor delay? A limp? Vision problems? Done. Done. Done.* But for some, those born on the very margins of viability at around 22 or 23 weeks, the cost in suffering can be very high, the prognosis deeply uncertain, and parents and doctors sometimes have to make terrible choices about whether to start or continue treatment. Even for us, safely and firmly four weeks past the viability zone, the doctors made it clear that they could not predict what costs Mira would bear.

There was nothing wrong with me, so I was discharged. I knew it was coming, but when I got in the elevator to go home, leaving her in the hospital, I lost my breath and bent double. My days crystallized into a strange routine. I pumped every three hours around the clock and froze most of what I pumped, since only the tiniest amount of milk was going down Mira's tube. I got up in the morning, pumped, and then drove from our apartment in Brooklyn to the NICU in Manhattan. Amol took two weeks off and then had to go back to work: Our medical insurance depended on it. I sat by Mira's pod. I watched the monitor that showed her pulse oxygenation, respiratory rate, and heart rate. I read *The Martian* by Andy Weir, which seemed appropriate, since I also felt stranded on another planet. Mira's lips were chapped from the oxygen mask over her nose and mouth. She often batted at it with her little red hands. The only time her wrinkled face relaxed was when a nurse removed the mask for a few seconds a couple times a day to make sure her skin was holding up, that she wasn't developing lesions. I'd take the opportunity to dip a piece of gauze in sterile water and rub it across her lips to moisten them and rub away the dead skin, and her whole body would go limp in what looked like relief.

Several times a day Mira's heart rate would suddenly plummet toward zero, setting off a round of increasingly shrill beeping from her monitor. A nurse would hustle over; pause a moment to see if Mira could handle the bradycardia, or low heart rate, episode on her own, and, if not, tap her back or chest to get her heart going again while I sat frozen, watching the number, willing it to climb. There was nothing wrong with her heart. The same thing would happen to the baby in the incubator across from ours, and it would be my turn to watch the back of the mother sitting there stiffen as she stared up at the number on her baby's monitor. Sometimes very premature babies forget to breathe. Their hearts neglect to beat. That is the kind of thing that is completely routine in the NICU.

I was allowed to do kangaroo care once a day, for up to three hours. Those were the only times I could hold her. I was not to cuddle or stroke or speak too loudly, which would overwhelm her delicate brain and could cause a bradycardia episode or lead to sensory problems when she was older. I cupped her tiny head and her tiny bottom against me and reclined, entirely still. I looked forward to those hours so much that it felt like a bad case of nerves before a date. And I started to lose my mind.

Maybe we were all a little unhinged, those of us keeping vigil by an incubator day after day. I remember one mother in the breast-milk pumping room who obsessed over her own bowel movements, plagued by the feeling that something unnameable and terrible was wrong with her. My thwarted mind started to behave strangely, unreliably, spitefully. A five-pound baby would come into the NICU and I'd think, *Jesus, what is that giant baby doing here?* Or someone's husband would cough a few times, spreading infection, I imagined, and I'd fantasize about slowly strangling him or cocking a gun and firing. It was a ferocious, helpless, wounded-animal response—*Stay away from my baby*—one that made me unrecognizable to myself.

Infection, including a cold, was one of the things that could, in theory at least, kill Mira. It was the middle of flu season. When I got to the NICU in the morning, I put on surgical gloves and pulled out the super-extra-sanitizing wipes that kill HIV and hepatitis C. They said FOR STAFF ONLY, but I thought they probably wouldn't kick me out for using the wrong wipes.

I claimed a chair and wiped it down. I wiped down my phone, my Kindle, my bag, especially the straps, and the surfaces around Mira's incubator. If someone, anyone, touched any part of my chair during the day, I'd wipe the whole thing down again. Before I came close to the incubator, I washed my hands with scalding water and then rubbed them with foaming hand sanitizer. If my hands touched anything—my jeans, a magazine, my face—I would sanitize them again. Before kangaroo care, I would run to the bathroom—using paper towels to avoid touching the door handle, the sink, any surface—and wash my chest, arms, and hands. Then, back by her pod, I'd rub my entire chest down with the foaming sanitizer, then frantically fan myself dry. At one point, one of my favorite nurses looked me up and down as I gobbed hand sanitizer between my breasts and asked, "Has your chest been somewhere I should know about?"

Doorknobs began to terrify me. Stores were full of danger—other people, multiple surfaces touched by so many. I'd shoot dagger eyes at anyone who came within a foot of me at a bodega. If you sneezed in my general direction, I genuinely contemplated murder. I'd bring groceries home and scrub them with sanitizing wipes. Who knew how many people had touched that can of tomatoes? Amol caught a cold and I was afraid to touch him or go anywhere near him. He slept on the couch. I demanded that he wear a surgical mask and gloves around the house until he was better, and maybe a little longer than that. My hands cracked and bled from all the washing, and I was secretly glad. It seemed appropriate; it was the way I felt inside made visible.

It's obvious to me now that I was experiencing postpartum anxiety or depression or both. But the situation seemed so extreme that it was hard to modulate my response. If a doctor tells me that, in theory, my baby could die of infection, is it reasonable to wash my hands twice? Ten times? Studies have shown that the parents of NICU babies are at risk for post-traumatic stress disorder (PTSD)— especially symptoms like fearful hyperarousal. One nurse I talked to said that she feels there should be a therapist for the parents on staff at every NICU. As it is, the nurses end up fulfilling that role as best they can.

Really, the nurses run the NICU. The physicians pop in and out, but it's the nurses who notice when something is wrong, who know when to recommend a blood transfusion, who restart babies' hearts dozens of times each day. For the smallest babies in our NICU, there was a ratio of two patients per nurse, so the relationship was intense. All day I watched the nurses. I eavesdropped on their conversations about where to get lunch; I imagined their lives. In a strange way I loved them, was obsessed with them. They were all women: fit, ponytailed, sneakered, and swift. They looked like Neutrogena commercials. They handled Mira and all her wires and probes with infinitely gentle skill, like someone wrapping the most fragile gift in the world or dismantling a bomb.

Every three hours our nurse would do Mira's "care": raise the lid on her incubator to change her diaper, take her temperature, check her skin for lesions, adjust her CPAP mask, and shift her position. They'd attach an empty syringe to the end of her feeding tube and draw up to check the contents of her stomach. If she'd successfully digested the last tiny dose of breast milk, another dose would be queued up to drip down into her over the course of the next several hours.

There was something about the open-ward setting, in which we were sitting inches from other parents and other babies, that paradoxically didn't encourage intimacy. Maybe it's that there was already too much of it. But there was one mom across from us whom I started chatting with in the early days. Her son was one of the only babies smaller than Mira in the NICU. She and I used to sit in companionable silence next to our babies, sometimes with our husbands, too, and then meet in the pumping room. We'd sit facing each other on the plastic chairs, boobs out, nipples suctioning in and out of the pump flanges, and make small talk: about her dog, our jobs, the logistics of taking maternity leave so early. How to get your milk to come in. (The NICU experience is not a recipe for successful lactation.) We'd ask about each other's babies, how we chose their names, how they were doing. She always said her son was critical but stable, but it was clear that he was sicker than Mira, with multiple organs not working on their own. She still hadn't been able to hold him. She never got more than a few drops of milk in those pumping sessions, but she always painstakingly saved them, and never stopped trying.

A couple of days before Christmas, I caught a cold. It meant I couldn't go see Mira, who was nearing one month old. I stayed home. I pumped. On Christmas Eve, Amol came home from the NICU looking gray. The baby boy next to Mira had died.

I never saw that woman again, but I think about her every day. I picture her in the pumping room, never giving up, saving the drops. I say her son's name to myself. I try to remember him well, his tiny form behind clear plastic, his parents sitting next to him.

The doctors never said Mira would be okay; they simply told us the news of the day, which, because we were lucky, was almost always cautiously optimistic: She was stable; she was growing. This narrowness of information was by design. Our favorite nurse told me later that the staff are careful never to say a baby is out of the woods until they are being carried out the door because babies have passed away days before discharge, struck by aliments like RSV, a common respiratory virus. So I hadn't even allowed myself to think of the possibility that Mira could ever come home when I got to the NICU one morning and found a flyer taped to her pod, instructing me that Amol and I would both need to take an infant CPR class before discharge. There we learned how to do chest compressions on little rubber dolls and all about SIDS (sudden infant death syndrome), which premature babies suffer from more than term babies. (Which seems like insult upon injury.)

The hurdles to cross before she could come home were routine for a baby her size, and yet seemed insurmountable. She needed to be at least four pounds, more than double her birth weight. She needed to be able to maintain her own body temperature and eat on her own. She needed to be able to breathe by herself without desaturating or pausing in an apnea. And her heart needed to keep beating on its own, without help.

These episodes, called bradys for short, still plagued her several times a day. Amol and I would call the NICU right before bed and first thing in the morning to find out if she had gained any weight and if she had had any bradys. The episodes could be triggered by stress or exertion. If she had a lot of episodes in a day, there was worry that it presaged something more serious—an infection, say, which thankfully never happened, or a need for a blood transfusion, since her bone marrow wasn't producing enough red blood cells, which did happen (another totally normal development). They could even be triggered by getting too comfortable, as when she was snuggled against my chest during kangaroo care.

One afternoon when Mira was about a month and a half old, we had a young nurse named Sarah. I liked her but—in my hyper-judgmental state—I thought she seemed more like a twenty-something you'd see at an annoying midtown bar than someone you'd want holding your baby's life in her hands. It was an unfair thought, but that's where I was. That day I was holding Mira when her heart rate started to plummet and the alarm pinged, first softly then more insistently. Sarah came over, observed the customary pause to see if Mira could figure this out on her own, and then started rubbing her back. Usually a rub, a pat, did the trick. This time her heart rate continued to fall. Sarah picked up Mira, who was totally limp, not breathing, and placed her on her back in the open incubator. She rubbed her chest more vigorously. Nothing. All the numbers on the monitor were falling, blinking, beeping. Mira's skin had turned gray, a color I have seen only once before, when my mother died. With one hand Sarah grabbed a nasal aspirator bulb and, with the other, reached for the ventilator bag that hung by every bedside, in case of an emergency need to breathe for the infant, or "bag" them. But first she stuck the nasal bulb up one tiny nostril and used it to suck out some snot. And just like that, Mira took a big breath, jerked. The numbers climbed; her skin pinkened. Sarah and I looked at each other and let out our own shuddering breaths.

As Mira's due date approached, she started to seem more like a baby. First, she got to wear clothes—a preemie-sized onesie that hung off her. After about six weeks on CPAP, she graduated to a nasal cannula that delivered oxygen, and then nothing at all. At around 33 weeks' gestational age, an occupational therapist taught us how to bottle-feed her, very slowly to avoid overwhelming her, which would inevitably lead to her forgetting to breathe and having an episode. After a few weeks of that, they took her feeding tube out. For the first time, nearly two months after her birth, we saw her face clearly, freed of the tubes and tape. One morning Amol and I came in and found her awake, her huge gray eyes looking around, as though she had just arrived—which, in a way, she had.

Finally, she hit four pounds. Now we were waiting for the bradys to go away. The policy was that she had to be episode-free for three days to go home, and if she had one, the clock was reset. She had gone two episode-free days when she had another one, while drinking a bottle. The attending neonatologist decided that, since the brady happened only because she got overwhelmed while eating, it didn't count. Presumably if her heart stopped while we were feeding her at home, we'd notice, as opposed to if it happened while she was sleeping, when we might not. I did not find this reassuring. I wanted to take her home, but I was terrified to take her home.

The next day, January 27, fifty-nine days after her birth and twenty days before her due date, Mira was unhooked for the first time from the monitor that tracked her heart rate, respiratory rate, and blood oxygenation, and we were told we could go home. It felt radical, like free-fall. It had snowed the day before, and I sat in the back seat with her, barely breathing myself, as Amol drove slowly down the icy FDR Drive to the Brooklyn–Battery Tunnel, all of us retracing our steps for the first time together since the Sunday morning in November when she was born.

OUR STORY IS like a kaleidoscope of lucky versus unlucky: The pattern shifts depending on how you look at it. We had excellent prenatal care; if we hadn't had that extra monitoring, Mira would likely have been stillborn. Only about 1.6 percent of babies are born before 32 weeks; only about 0.5 percent are born weighing less than 1,000 grams. We had made it four weeks past what anyone would consider the viability line. But being IUGR increased her risk of death or disability to the equivalent of babies born two to three weeks earlier, 25 or 26 weeks. She was born in a hospital with a level four NICU, the best. Ours was one of the many preterm births for which there was no answer as to why the problem happened—no risk factors, no health problems, no reason.

The incredible truth is that, from a medical point of view though not from a parental point of view, Mira's NICU course was unremarkable, at least once she was stabilized and started to gain weight. As recently as the 1970s, she might have been simply allowed to die in the delivery room, since she was under 1,000 grams. But in the context of today's high-resource NICUs, Mira *was* relatively unremarkable. She was your average one-and-a-half-pound baby whose heart stopped several times a day, who needed blood transfusions to stay alive, who needed life support to breathe, eat, and stay warm.

When Mira was born, I felt uniquely bad at gestation, when I was actually one of almost a half million mothers who would give birth early in the United States that year. I didn't know that our family's private gauntlet was a small part of a worldwide public health emergency.

For most of human history, gestation was a solo affair, and it either resulted in a living baby and surviving mother or, lots of times, it didn't. Left to nature, most preterm babies die. In today's NICUs, most of them survive.

Throughout the middle and latter half of the twentieth century, extraordinarily headstrong doctors and scientists were willing to risk their reputations to argue that something more could and should be done for the tiniest babies. As a result, in two generations, we've gone from having essentially no treatment for premature babies to flirting with viability just past the midpoint of pregnancy.

It's changed the way we think about babies and our obligations to them. It's changed the ways we understand what it means to be alive, what it means to be human, and what constitutes a life worth living. At least for the moment, it messily defines abortion law in the United States, since "viability" is determined by neonatology's success. (As I write this, several states have passed clearly unconstitutional laws that essentially outlaw abortion, with the goal of getting the Supreme Court to reconsider and potentially overturn *Roe v. Wade*, which established the right to an abortion up to viability.) Premature birth—and which groups suffer from it the most— reflects back all the deep racial and economic injustices that plague the United States.

And sometimes, the project of treating premature babies reflects what is best and most beautiful in all of us: that we will go so far for a single human life; that some people dedicate their lives to the children of others; that love of all kinds can burn so bright.

Amol likes to think of Mira having a superhero origin story: She battled the machines and lived. Now she carries hidden power. I sometimes think of it as a fairy tale—the old, dark kind—a girl who had to find her way through a dark wood full of sharp teeth, alone and brave before her time. I think we are both trying to describe how her birth and her struggle feel fundamental: at the intersections of birth and loss, science and humanity.

Part I

The Unexpected

Millions of Births

1

What Happened?

ABOUT A YEAR and a half after Mira's birth, when she finally started sleeping through the night, I started to emerge from the deep, consuming fog of anxiety that had enveloped me through her babyhood. After she was born, I was engaged in a balancing act of survival, so afraid to look down that I took in only as much information as was offered.

As I started to come back to myself, to the new landscape that was our family, I had a question: *What happened?* For us, as for many families, there is no answer to that question. No one knows why the placenta failed; the lab tests were inconclusive. No one can tell me why Mira had to come into the world the way that she did. Or, for that matter, why I did, too.

So I started to ask the same question in a bigger sense: *What happens?* What happens to one in ten babies and their families?

When I first heard Mira might come a little early, my conception of preterm birth was . . . well, embryonic, even though I had known the basic gist of my own early arrival all my life. As I peered back into my memories of her first months, this subject opened itself up in strange and profound and confusing ways. More questions suggested themselves: the weird intersections that had fluttered by, the science I had half understood, the other families that we had passed in that fog. The astounding fact that babies like Mira are now more than likely to survive. *What happened?*

IF YOU'VE BEEN pregnant, you probably know the prenatal visit drill, and it goes something like this: risk of Down syndrome, diet and weight gain, Pitocin and epidurals and C-sections, genetic testing. There are certain topics that come up over and over in routine obstetric visits. Preterm birth is usually not one of them.

I remember, in particular, how much emphasis was put on finding out if my fetus had Down syndrome, since I was going to be thirty-five when I gave birth, putting me over the line into "geriatric." But as a healthy, nonsmoking, thirty-five-year-old, middle-income white woman living in Kings County, New York, I had about an 8 percent chance of a preterm birth. In other words, even *without* the extra risk indicated by the low PAPP-A, I was already *29 times more likely* to give birth early than I was to have a baby with Down syndrome. Chromosomal conditions like Down syndrome can't be cured, of course, but once they are discovered, there is a decision to be made: continue the pregnancy or end it. And physicians like to offer medical options; I don't blame them. With preterm birth, this is impossible. There is precious little scientific consensus on why it happens, much less how to prevent it. In many cases, there is nothing much to *do* but wait, watch, and hope to go full term. It is a phenomenon defined by uncertainty. And to complicate matters further, it is a pregnancy outcome with many different causes, not a single disease or condition. Any true statement about prematurity has at least a dozen caveats.

ONE OF THE most astonishing facts about preterm birth is just how common it is, how devastating it is to the health of women and babies, and yet how invisible. Premature birth is the leading cause of newborn death in the United States and worldwide. It has surpassed pneumonia as the leading cause of death for children under five throughout the world. It is also a leading cause of developmental problems of all kinds, learning disabilities, and chronic conditions like asthma.

Prematurity costs the United States about $26 billion per year, and, more importantly, it is the cause of untold sorrow and struggle. The American preterm birth rate is 10.02 percent and rising; that's almost 400,000 premature babies each year—the worst rate in the industrialized world. Worldwide, 15 million babies are born too soon, and about 1 million of them die as a result.

The consequences of a preterm birth can follow even the luckiest of families for years, sometimes for entire lives. It is a postpartum anxiety or depression that is too slow to ebb, a lingering vigilance, an alarm that keeps ringing in your head. It is dragging an oxygen tank with you to the playground. It is carting a baby, then a toddler, then a preschooler, to physical and occupational therapists and learning specialists and pediatric pulmonologists and neurologists. And yet, experiencing premature birth can also forge a sense of wonder, gratitude, and strength.

WHEN MIRA WAS in the NICU, I was terrified. When she got home and I found out that she needed physical therapy for a motor delay, I was overwhelmed by the very notion of navigating the Early Intervention System and angry that I had to think about it. When she had her first asthma attack—a common complication of prematurity—I held the nebulizing mask to her face, felt her little rib cage contracting with each labored breath, and wanted to scream.

But also, being her mother is the greatest joy of my life. When she was a baby, I couldn't get enough of the way she kick-kick-kicked her little legs when she was delighted, the hugeness of her grin, the grave look on her face when she studied a stuffed animal. As she grew, she became physically fearless, rough-and-tumble, a tiny, delicate-looking child who was anything but delicate. She is sharp, inquisitive, funny, and thoughtful. She is formidable. She has amazed me from the moment she was born.

For the richest country in the world, the United States' preterm birth rate is clearly too high, and so is its sad corollary, infant mortality. The fact that we do have such a high rate, and that it's even higher for Black women and other marginalized communities, is an indictment. Cynthia Pellegrini, a senior vice president for public policy and government affairs at the March of Dimes, compares preterm birth rates to the canary in the coal mine. "If those numbers are going up," she said, "it means that something is wrong." And the numbers *are* going up. For a resourceful, rich country that claims to value families, having a child is fraught with avoidable suffering and risk.

However, as always, statistics don't tell the whole story. Sometimes an early birth is the best possible outcome: In the absence of other treatments, it might be necessary to avoid the risk of stillbirth, as it was for Mira. So a rise in preterm birth can also mean a decrease in stillbirths. Or an early delivery might be necessary to protect a mother's life, for instance, when she has preeclampsia. At a March of Dimes conference that I attended in 2018, a physician made this key point: For a patient who has given birth previously at, say, 26 weeks, a second pregnancy lasting 35 weeks is a triumph even though it's still classified as a preterm birth. Conversely, if a woman presents with preeclampsia at 36 weeks, delaying delivery long enough to push her over the term line to 37 weeks at the expense of her or her baby's health is not a success—although it might look like one statistically.

IT HELPS TO understand the basics. A typical healthy pregnancy lasts between 37 and 42 weeks. Any birth before 37 weeks is premature, or preterm. Gestational age is simply the number of weeks since the start of the pregnant person's last period; most people conceive around week 2. Preterm babies continue to go by gestational age, even outside the womb, until they hit 40 weeks. In other words, when Mira was one month old, she could have been called a 28-week baby who was now 32 weeks' gestational age.

Broadly, there are two kinds of preterm births: those that are the result of spontaneous preterm labor or rupture of membranes (water breaking)—which account for about two-thirds of all preterm babies—and medically induced early deliveries (either C-section or vaginal), which make up the remainder. When providers decide to induce a premature birth, it is because something has gone wrong, and birth is necessary to protect the baby's life, the mother's life, or both.

There are lots of different causes for both kinds of preterm birth, but many preterm births don't have a known medical explanation. When parents ask, *What happened?* the only answer is, *We don't know.*

However, there are some established risk factors: High blood pressure, either previously existing or pregnancy related, can lead to one of the most common causes of premature birth, preeclampsia. This is a condition that causes damage to maternal organs like the liver and kidneys, and it can escalate into a life-threatening syndrome if left untreated. Preeclampsia often necessitates delivery to protect both the mother and baby.

Placental problems, like a placenta that stops working unexpectedly, as mine did, or one that abrupts (partially separates from the uterine wall) or is attached to the cervix (placenta previa), can lead to premature labor or to a medically indicated premature birth. It can also happen because of a cervix that won't stay closed, or an irregularly shaped uterus.

Infections, like those of the urinary tract or the amniotic fluid and membranes, can trigger premature labor, especially when they go untreated.

Women pregnant with multiples are more likely to give birth early, as are women who used in vitro fertilization (IVF) to get pregnant and women younger than twenty or older than forty. Being underweight or obese is associated with premature birth, as are behaviors like cigarette smoking. Exposure to clinically significant chronic stress, like income insecurity, racial discrimination, unsafe working conditions, and trauma, also puts women at higher risk of giving birth early.

Having pregnancies that are spaced closely together increases risk for preterm birth; experts advise getting pregnant again no sooner than eighteen months after giving birth. (After a first-trimester miscarriage or abortion, there is apparently no medical need to wait longer than one cycle to get pregnant again.) That means that access to family planning and contraception is a meaningful intervention against premature birth.

And people who have given birth early before are much more likely to give birth early again. The good news is that increased monitoring can help; your provider might suggest weekly progesterone shots if you've had unexplained early labor in the past, or low-dose aspirin if you have high blood pressure.

But there has not been as much progress toward the prediction and prevention of premature birth as there has been on the neonatal side, caring for preemies. At the moment, progesterone and low-dose aspirin are really the only two drugs available to stave off premature birth, and their effectiveness is limited. But a recent groundbreaking study from the University of California, San Francisco, used a blood test early in pregnancy to measure certain proteins and biomarkers. By evaluating the test results along with information like maternal age and income, researchers were able to predict whether a pregnant individual would go on to have a preterm birth with 80 percent or better accuracy. This would be extremely good to know—increased monitoring can improve outcomes significantly—but there would still be no sure way to stop it from happening altogether. At least, not yet.

Once labor starts in earnest, there are medications called tocolytics to slow or stop labor and delay birth.

These are mostly temporary measures, but a delay can be lifesaving, because one of the most important advances of the last several decades has been the use of corticosteroids, which became standard practice in 1994. When obstetricians think a baby might be born before 34 weeks, they give the mother a series of two of these steroid shots, twenty-four hours apart. The steroids rev up the fetus's lung development, which is one of the last systems to mature. This treatment makes a huge difference in both survival and in long-term health outcomes, but it needs at least twenty-four hours to take effect.

Once a baby is born early, the severity of their immaturity is categorized by gestational age. Extremely preterm is fewer than

28 weeks; very preterm is fewer than 32 weeks; moderately preterm is fewer than 34 weeks; and late preterm is fewer than 37 weeks.

Some providers have proposed a separate category: Babies born at 25 weeks or younger are sometimes called profoundly or severely preterm, or "periviable," for the way they straddle the viability zone.

Birth weight is divided into similar categories: Extremely low birth weight is under 1,000 grams (2.2 pounds); very low birth weight is under 1,500 grams (3.3 pounds); low birth weight is under 2,500 grams (5.5 pounds).

Often the immaturity and weight categories track together: extremely preterm babies are usually extremely low-birth-weight babies. And those categories are sometimes lumped together in studies—for example, a study might include outcomes for both extremely premature babies and babies under 1,000 grams. (Birth weight is a more precise measurement than gestational age.) But because some babies are growth restricted, or small for gestational age, or the opposite, the groups are not always equivalents. Mira, for example, was very preterm but extremely low birth weight. She was more mature than her birth weight would suggest.

A neonatal intensive care unit is a special nursery where very premature and critically sick babies can get all the medical care they need, including incubation, life support, and sometimes surgery. Premature babies are not the only babies who need NICU care—term babies can have complications, disorders, and illnesses—but generally premature babies make up the majority of NICU patients.

2

Treatments and Outcomes

EVEN IN THE absence of other complications and illnesses, premature babies generally have trouble with three crucial abilities: staying warm, breathing, and eating. (Before about 34 weeks, babies can't coordinate sucking, swallowing, and breathing at the same time.) Of these, breathing is the most dangerous deficit, although respiratory support has improved tremendously in the last generation.

All of these basic problems can be mitigated by life supports. Incubators give warmth and humidity; feeding tubes deliver breast milk or formula. Respiratory support is generally provided by positive-pressure ventilators, which require intubation; CPAP machines, which offer less invasive, non-intubated ventilation; or nasal cannulas.

Treatment starts in the delivery room: Some infants don't breathe after birth—the lower the gestation, the more likely this is—and must be resuscitated immediately, which often means putting them on a ventilator of some sort, but can also include other, more intensive measures, like chest compressions and adrenaline.

The two most significant outcomes measured for preemies are mortality—usually defined as death before discharge from the hospital or death in the first year of life—and morbidity, which describes any illness or disability (short- or long-term) resulting from prematurity. Generally, the earlier the gestation, the higher the rates of both mortality and morbidity.

In the shorthand of talking about premature babies, gestational age tends to be the most-used description (like "26-week infant," "32-week infant"), because the degree of immaturity is usually the most important factor in outcome. But a particular baby's risk of both mortality and morbidity is idiosyncratic and dependent not just on length of gestation but also on birth weight (the bigger the better), sex (girls do better), steroid exposure (crucial for lung and brain health), whether or not they are one of multiples (singletons do better), and the reason for the birth. Babies born with an infection or other health complications tend to do worse than babies born as a result of spontaneous labor to a healthy mother. So discussing babies of a certain gestation as a monolith is common but also a bit misleading: One 24-week baby might be so different from another 24-week baby that equating them doesn't make sense.

At 28 weeks and 845 grams, we found ourselves at a lucky crossroads: During my C-section, a nurse said to me, "Nurses always breathe a sigh of relief when we get to 28 weeks." I didn't really follow her logic at that moment, but what she meant was that 28 weeks is a kind of dividing line of severity, a gestation that is not at all ideal but at which survival starts to be extremely good and long-term disability trends toward the minor.

The CDC collects infant mortality statistics that give a blunt, population-level look at death by gestational age at birth. (Again, these statistics don't exactly correspond to any individual baby's chances, which depend on individual circumstances.) Those data say that, in 2016, 11 percent of all 22-week infants, 43 percent of 23-week infants, 65 percent of 24-week infants, 78 percent of

25-week infants, and 86 percent of 26-week infants survived the first year of life. At gestations greater than 27 weeks, survival was 90 percent or better.

There are several main reasons premature babies die, but most deaths really come down to one thing: immaturity. Their little bodies are not ready to be here. Specifically, that might manifest in severe cardiovascular and lung problems as a result of respiratory distress syndrome (RDS); intraventricular hemorrhage, or bleeding in the brain; necrotizing enterocolitis (NEC), in which a portion of the intestine dies; and infection, which they might be born with or they might acquire in the hospital. Some babies who experience these complications survive, but others do not.

Some deaths, including many of the deaths at 22 weeks in particular, are the result of letting the babies die—not initiating active treatment at all, but opting for palliative comfort care in the delivery room. Around 60 percent of 22-week babies do not get active treatment, without which they inevitably die. When the chances of survival are extraordinarily slim, it is sometimes the kindest course—intensive care can be painful and burdensome for the infant and can be justified only if there is a reasonable likelihood that they can benefit from it.

Disabilities that result from a premature birth are classified as mild, moderate, or severe, but even within those categories, actual experiences are quite diverse.

One of the most common causes of developmental disability in kids born early is cerebral palsy (CP). This is a motor disability caused by a brain injury to the fetus or newborn baby; the injury might be the result of lack of oxygen to the brain at birth or of bleeding in the brain. Sometimes the injury also causes some intellectual disability. Mild CP might manifest as a child who is almost imperceptibly clumsy or one with tight muscles on one side or who has a limp; moderate CP might manifest as a child who uses braces to walk; severe CP might manifest as a child who uses a wheelchair. CP is permanent, but functioning can greatly improve with various therapies.

General developmental delays and issues—like a child who is late to speak, late to walk, or has difficulty eating or with sensory processing—are relatively common in preterm babies, and while they may require intervention and therapy of some sort, they often resolve or at least improve. Some disability statistics are based on follow-up at one to two years of age, and some of the children who are delayed at that point will grow out of it, which is certainly not to suggest that it is easy on them or their families.

In fact, a delay is built into the assessment of babies and children who were born early. Until the age of two, preemies have two ages: an actual age, counted from their birthdays, and an adjusted age, or the age they would have been if they had been born on their due date. If they are developing typically according to their adjusted age, they aren't considered clinically delayed until they hit two years old.

When Mira was first assessed by Early Intervention staff at about four months old actual and one month old adjusted, her test scores would have indicated a severe developmental disability if she had truly been a four-month-old. In fact, they reflected only a minor delay when tracked with her adjusted age. Adjusted age is very real: The body has its own timeline, and birth doesn't accelerate it. It can actually be the opposite: When birth happens too early, being out in the world too soon can slow development down.

All these idiosyncrasies make it hard to give exact rates of disability by gestational age at birth. A further complication is that one of the biggest, best data sets that exists on extremely preterm outcomes, from the National Institutes of Health's Eunice Kennedy Shriver National Institute of Child Health and Human Development's Neonatal Research Network, relies on data from 1998 to 2003. Presumably outcomes have gotten better, but we don't really know for sure.

The vast majority of preterm babies grow up to have no disability or a minor disability. But the likelihood of a disability and the likelihood that the disability will be serious increase as gestational age at birth decreases. At 22 weeks, for instance, according to some data—and again, this varies by circumstance—75 percent of survivors have either a moderate or major disability, defined as moderate to severe cerebral palsy, blindness, or deafness. By 25 weeks, about 39 percent of survivors have a moderate or major disability.

When I think back on what the neonatology fellow told us about Mira having a 50 percent chance of having a disability, based on what I know now, that sounds high—although (like everything!) it depends on how you count it. If you count all developmental delays, even those that don't substantially impact functioning in the long run, it's probably about right. But I wish he had been more specific about what he meant. And it turns out that Mira does have ongoing motor delays that technically classify her as having a disability; she has to work harder to master certain tasks than her peers do. However, to an outside observer, the difficulty is almost imperceptible.

I have found statistics useful and interesting in trying to understand premature birth in a big-picture way. But in my own life with Mira, I found statistics to be fairly unrevealing and more likely to make me feel panicky than informed. Knowing that your child has a 50 percent chance of disability isn't helpful, especially if no one ever makes it clear exactly what that means.

So if you are looking for answers, please don't despair over statistics, which may not be truly relevant to your child. Ask your providers about your particular child, and keep asking until you understand what they really mean.

3

Viability and the Zone
of Parental Discretion

THE WORD "VIABLE" simply describes the ability of a fetus to survive outside the womb. For parents, it can be an anguished hope; for neonatologists, a Hail Mary try: Can this baby live? Before respiratory support, it was unusual for a baby under 30 weeks gestation to survive. Now, with the most advanced care and in the most fortunate situations, it is possible at 22 weeks. But viability is culture- and resource-specific. In countries without advanced life support, it is still rare for very preterm babies to live because they don't have the battalion of treatments that makes survival possible. In industrialized countries, practices can vary by culture. In Japan and Sweden, it is common to attempt to resuscitate and treat 22-week babies. In the Netherlands and France, babies under 24 weeks are rarely resuscitated—not because these countries don't have the resources, but because it's not considered the right thing to do. "When is viability?" is a question with many answers.

For about the last thirty years, 24 weeks was widely considered the limit of viability, meaning babies born at fewer than 24 weeks' gestation were thought to not have a reasonable chance of survival even with the most sophisticated treatment, and were mostly allowed to die if born alive. But in recent years it has become quite common to attempt treatment for 23-week babies, with some success. Some hospitals will try to treat babies as young as 22 weeks, especially if the parents request it, or even very, very occasionally 21 weeks if there is significant uncertainty about gestational age. In fact, a Swedish study recently documented that, between 2014 and 2016, nine 21-week babies were admitted to a NICU and two of them survived.

Neonatologists generally talk about viability as a zone or a period of time, sometimes a "period of periviability," or a "zone of parental discretion," or a "gray zone." Basically, it is the period of time in which a baby may be too immature to survive but may not be. Just about everyone agrees that this period of time is one in which it is medically reasonable to choose either active treatment or comfort care—essentially, having parents hold a newborn until the baby dies. Therefore, within that time of deep medical uncertainty, many ethicists agree that the choice between active or comfort care should be made primarily by the family, with input and support from doctors. Not everyone agrees on where the boundaries of that zone are, but they are roughly 22 to 24 weeks.

For those born approximately 25 weeks and above, the baby's chances of survival and good health are so great that NICU care is usually mandatory. For example, if I had asked that my 28-week baby not be resuscitated—and very occasionally this actually happens—the hospital would have done it anyway. On the other hand, if I had delivered at 20 weeks and insisted I wanted my baby intubated and taken to the NICU, as far as I can tell, there is not a hospital in the world that could or would carry out that request.

Viability is important for other reasons, too. Some legal traditions use viability as a kind of proxy for the fetus acquiring some of the protections of personhood. In the United States, for instance, under *Roe v. Wade*, state governments may prohibit abortion after viability. But viability is not necessarily a clarifying distinction. Viability has as much to do with the medical care available as it has to do with the baby. If viability equals personhood, for one thing it would mean that babies in rich countries become people before babies in poor countries do. If you are looking for a bright and certain line, an objective criterion, well, viability is not it.

In the industrialized world, the viability line has shifted over time as the care has improved: In the early 1970s, at the time of the *Roe v. Wade* ruling, babies around 28 weeks and up were considered viable. Over the following decades, with advances in technology and medicines, that line trended younger, and therefore, so did some states' limits on abortion. And this is how neonatology and abortion law are yoked to each other.

Even absent the needlessly complicating political overlay of abortion law, making decisions for premature babies within the viability zone is extraordinarily difficult and painful. Within that periviable threshold, there is little consensus on what is best for babies and families—what is most humane and ethical or even legal. Over fifty years after the advent of modern NICUs, parents and providers still grapple with these questions every day. It's a dance in the dark.

Part II

The Body

Incubation

4

The History of Incubation

Coney Island, Chicken Eggs, and Changelings

There came into the world a child . . . who was colourless, sightless, voiceless, and so poor a weakling that all despaired of him except his mother.

—VICTOR HUGO, WRITING OF HIMSELF, BORN PREMATURELY IN 1802

LUNA PARK, CONEY ISLAND, 1903. If you were to stroll down the boardwalk, past the gondola rides, the monkey theater, the clowns, the high-wire acts, the "Hindoo village," and the wild beast arena, you might have eventually found yourself in front of an ornate edifice with a ticket booth out front. And, in lights: "Infant incubators with live infants. All the world loves a baby!" Entry was 25 cents, the equivalent of 7 dollars today. Inside, the hustle and razzle-dazzle of the boardwalk vanished: Here was a spic-and-span medical clinic staffed with nurses in starched white dresses and caps. And, behind guardrails, rows of tall, glass-fronted incubators, each with a tiny baby inside. They were dressed in doll's clothes, or sometimes in regular baby clothes that were far too big, meant to emphasize their smallness. It was a sideshow, and it was also America's first neonatal intensive care unit.

Martin Couney, the bow-tied, mustachioed gentleman running the incubator show, was a huckster, a cipher, a genius. He had been born Michael Cohn in 1869 in what was then Prussia and is now Poland, and immigrated to New York at the age of nineteen. As for the crowds clustered at the guardrails, peering at the babies, they were just reacting the way that people have always reacted to premature infants: with a mixture of tenderness and horror, awe and discomfort.

At the time, there were virtually no formal treatment options for premature babies in the United States, and they usually died. The incubator had been invented in France twenty-three years earlier, but it was still mostly unknown in the United States. When a very small baby was born, the midwife or obstetrician, if there was one, usually told parents to expect their child to pass in a matter of hours or days. When word of Couney's success spread, families started bringing their tiny newborns to Luna Park because they heard he might help, and they had nowhere else to turn. Meanwhile, parkgoers couldn't get enough of the drama, the cuteness, the weirdness. Some fans came every week to check on a particular baby's progress. It was like a prototype reality show, one with excruciatingly high stakes. *Live infants! All the world loves a baby.*

The story of incubation is rich and strange and full of detours. It is not a straightforward tale of medical innovation and progress. Gestational technology, from the simplest incubator to a science-fictional artificial womb, has always inspired anxiety about motherhood and personhood, about what is "natural," and about who deserves to be here. The act of gestating a baby outside a mother brings up complicated feelings about what it means to be human, and whether some of that humanness can be lost.

Premature birth has not always been regarded as a medical problem with medical solutions, but early birth has always happened. Sir Isaac Newton, Albert Einstein, and Mark Twain were supposedly all preemies. And prematurity has long had a place in the popular imagination: In *Macbeth*, the title character is killed by Macduff, who was "from his mother's womb untimely ripped," validating the witches' prophecy that Macbeth would never be harmed by a man born of a woman. In other words, because Macduff was born early by C-section, he wasn't truly birthed by his mother. He wasn't a man like other men. Macduff's untimely birth makes him superhuman, otherworldly, the answer to a riddle.

For most of human history, babies belonged to the home, to women. Even when doctors started attending some births, in the nineteenth century in parts of Europe and the United States, the fetus itself was a mystery, a total unknown, at least until it was born and either lived or died. Mothers often had a sense of how many months they had been pregnant (it was sometimes especially important to be able to prove the baby was legitimate) and may have known when a baby came far too early. However, gestational age was a fuzzy concept to say the least. A French textbook from 1683 posits that a child born before completing seven months of gestation can't live because of its "weakness," which was a common way of describing the constellation of problems that prematurity brings.

Women and community midwives carried knowledge and practical expertise about childbirth and infant care; in many cultures there were midwifery techniques for babies who didn't breathe right away: slapping the baby, swinging the baby back and forth, bleeding the baby, bathing it in brandy, and doing mouth-to-mouth resuscitation.

Reactions to the birth of a weak, tiny baby vary across time and culture, but there are common instincts, and one of them is devising ways to keep the baby warm—incubating them before incubators existed. One report from 1913 described how members of the Thonga people of southern Africa would swaddle too-small babies in leaves from the castor oil plant and then put them in a pot in the sun. Another account described how Arctic Yupik people would take the skin of a seabird, turn it inside out, nestle a premature baby in the feathers, tie it into a bundle, and hang it over a lamp.

In my own family lore, my grandfather and his twin, who were born at home in 1921 to recent Italian immigrants in Boston, were each so small that they could be held in the palm of my great-grandmother's hand. She later said they were about one and a half pounds each. My great-grandfather thought they should call the undertaker, but my great-grandmother bundled them into shoeboxes packed with cotton and put them in a very low oven. They both survived in good health.

Prior to the invention of incubation, it was rare but not unprecedented for an early physician to take an interest in the treatment of an extremely tiny baby.

On April 19, 1815, a woman in Paisley, Scotland—we do not know her name—went into labor, and she knew it was far too soon for her baby to come. This was her sixth child. She thought that she had thrown herself into early labor because she had worked too hard the day before. She sent for Dr. John Rodman, who arrived in time to help her deliver the baby boy. Rodman quickly wiped the boy down and wrapped his entire body tightly in layers of flannel, leaving only his nose and mouth exposed. His mother took him straight into the bed with her and stayed with him there for two months, taking turns with two other women, possibly relatives, who were willing to lend their body heat.

She fed him breast milk by the drop, as well as bread boiled in sugar water and then strained through a linen to make a very thin gruel. Rodman suggested she add port wine to that gruel, as well as daily beef broth and castor oil for constipation. Rodman noted that after taking his port wine, the baby was "uncommonly happy." When they finally weighed the boy at three weeks old, he was an astonishing 820 grams (1.8 pounds). Rodman's account of the birth, which he found extraordinary—he calls the boy's health "peculiarly regular"—ends when the boy is four months old and able to be carried outside for short periods. It would have been unheard-of for a baby that small to survive at that time, but clearly that didn't stop parents from trying—although obviously not every mother would have had the option of staying in bed to personally incubate her child for two months.

But it was not always desirable or possible to express and prioritize parental love the way we do now. In ancient Sparta and Rome, babies who were born too small were often left outside to die of exposure. In Europe in the Middle Ages, up to half of babies did not survive infancy, which by some accounts led to a kind of "tenderness taboo," a bulwark against becoming attached when the child was just as likely to die as to survive.

Premature babies in particular have always inspired a certain ambivalence about whether or not they are "supposed" to be here—how much claim they have to resources, to life. This idea has not gone away; it plays out even now in subtle ways. (For instance: Much is made of how expensive NICU care is, when in fact intensive care for infants is many times more cost effective than intensive care for adults, in terms of producing healthy survivors.) In earlier times the birth of a small, weak baby was very bad news indeed: The time and attention demanded by a baby who would likely die anyway threatened a mother's resources for her other children as well as for herself. All over the world, folk practices emerged to help deal with this situation, sometimes by finding ways to tacitly (and deniably) sanction infanticide.

One of the best-known examples is the idea of the changeling, common throughout Europe and in parts of Africa. They were variously described as wizened, inconsolable babies, sometimes with other features like oversized heads and lots of body hair, characteristics that are typical of preterm babies. (Changelings were also described in ways that match up with congenital disorders.)

These babies, it was said, were not human at all but monstrous replicas left by fairies, spirits, or devils. In some traditions this meant that your real baby had been stolen away to the spirit realm. In response, you might leave the changeling in the woods for the night or bathe her in ice-cold water or put her in an oven—not so low this time.

It is hard to square these accounts with stories of people doing anything—everything—to keep their too-tiny newborns warm, but this contradiction is inextricable from the history of prematurity. It has been a long and winding road from leaving underweight babies to die in the woods to engineering $30,000 machines to cradle them.

Modern incubators—and, in a way, our modern concept of the premature baby—owe their existence to Dr. Stéphane Tarnier and his fateful jaunt through the Paris zoo in 1878.

Tarnier was born in a village near Dijon in eastern France in 1828. His father was a country doctor, and Tarnier followed in his footsteps, moving to Paris to study and practice medicine. In 1856 he started working at l'Hospice de la Maternité, where he was startled to learn that 10 to 20 percent of the women who delivered their babies there—women who were mainly impoverished—died of puerperal fever, a bacterial infection of the reproductive tract. There were weeks when the fever raged through the hospital and killed dozens of postpartum women at a time. The conventional wisdom held that somehow a lack of fresh air caused the fever.

Tarnier proposed that the rampant infections might have *something* to do with contagion: for example, the fact that doctors would do autopsies and then come deliver a baby without stopping to wash their hands, and the fact that women were packed three to a bed. Perhaps the fever was caused by something invisible that could pass from one person to another. This was a novel idea at the time, but the chief of the hospital was impressed and put Tarnier in charge of making changes: He quarantined those who fell ill, and he instituted brand-new ideas about antisepsis, including hand-washing and the use of antiseptic spray. Deaths from the fever plummeted.

And so, as the years ticked by, Tarnier rose to be surgeon in chief of La Maternité, a respected member of this new field of medicine, modern obstetrics. He was a disrupter before the term existed, a self-promotor. He wore jet-black three-piece suits that stretched over his stomach and had a jaunty goatee.

Who knows why he went to the zoo that day in 1878? He apparently knew the zoo's director, Odile Martin, but the exact reason that Tarnier wandered into the bird exhibit is lost to time. He paused in front of an egg incubator. Some say it was a chicken egg warmer, and others say it was an exotic bird egg warmer. Either way, it was a closed box heated by a hot-water tank on the bottom. This kind of apparatus for eggs had been used since the ancient

Egyptians invented it, but, looking at it that day, Tarnier wondered what would happen if he put a baby inside.

Tarnier had long been bothered by the problem of the smallest babies born at La Maternité: They died in large numbers, often of hypothermia. They were horribly skinny and their metabolisms didn't generate enough warmth to keep their temperature in a normal range, even when wrapped in fleece. At the time, babies born undersized or sickly were known as *débiles*, or weaklings, and no distinction was made between those with an illness or disability of some sort and those born too soon.

Tarnier asked his friend Martin to rig up a baby warmer for him, and in 1880 or 1881 he started using the prototype at La Maternité.

This first enclosed incubator was a wooden box with thick, sawdust-insulated walls and a glass lid so the infant could be monitored. The lower half was a hot-water tank, keeping the air inside hot and humid, around 90 degrees—a big improvement on the room temperature, which was often quite chilly. In the prototype, the top half had room for four babies, the same way an egg warmer cradles multiple eggs at a time. In later incarnations, each incubator was single occupancy and heated with water bottles on the bottom, which had to be refilled every couple of hours.

Dr. Jeffrey P. Baker, a Duke University professor of pediatrics who wrote a history of the incubator called *The Machine in the Nursery: Incubator Technology and the Origins of Newborn Intensive Care*, points out that Tarnier's apparatus was barely an improvement on what mothers had been rigging up for centuries. And simple warming bassinets heated with water had been used in maternity hospital settings before, although Tarnier's tightly closed top seems to have been a new idea, keeping the temperature more constant.

But Tarnier's real accomplishment was bringing the notion of warming tiny babies into the realm of formal medicine: He and his interns measured and published the results, giving incubation the kind of respectability medical people care about. And his data were exciting: Previously, 35 percent of babies born at La Maternité weighing under 2,000 grams (4.4 pounds) had survived. With Tarnier's incubators, 62 percent of them survived.

In 1884, Tarnier started feeding the babies by gavage, a tube down the throat through which small doses of breast milk could be poured directly into the stomach. (This technique is still used today.) Previously, babies who couldn't suck had been fed by slowly, slowly dribbling milk (or wine or gruel, et cetera) into their mouths or noses using a spoon or even a quill. No one seems to know where Tarnier got the idea for tube feeding, but it is plausible that it was another moment of inspiration via poultry: The practice of using gavage to overfeed geese and ducks so that their livers fatten up for pâté has been common for centuries, and might have been familiar to Tarnier from the French countryside.

With the combination of these strategies, Tarnier claimed to have successfully treated babies as young as 24 weeks' gestation, although he did not routinely put babies that small into the incubator, usually using a cutoff of about 2.5 pounds. (Birth weight is, even now, more precisely measurable than gestational age, which is why doctors have always relied on it.) Tarnier was chuffed. One of his students presented their results this way: "It is thanks to the use of the incubator and force-feeding [gavage] that we have managed, in recent years, to raise children who were not older than six months or six months and a few days of life intra-uterine. The results obtained are therefore remarkable." This is perhaps the first time a scientist claimed to have moved viability using a new technology. A previously immovable, natural biological marker was becoming a changeable technological marker. Viability was becoming a debatable idea.

Through the use of these interventions, a patient group started to take shape. Who was helped by the incubator and gavage? Babies born too small, too soon: premature babies. Baker, the pediatrician-historian, writes of it this way: "Tarnier and his co-workers constructed the concept of the premature infant even as they constructed the incubator."

Tarnier didn't have his eureka moment in a vacuum: There was tremendous anxiety in France about declining birth rates as the French population fell behind that of Austria-Hungary and Germany for the first time. The Franco-Prussian War of 1870 had ended with a brutal months-long siege of Paris by the German army, a humiliation that set off a political obsession with the strength and size of the French nation. Obviously, mothers were to blame! Women should have more children; it was for the good of the republic.

The zeitgeist was ripe for saving babies, and for technological optimism. (Other inventions of the late 1800s include the telephone, the electric lightbulb, the thermostat, the skyscraper, photo film, the ballpoint pen, the smoke detector, the zipper, and the machine gun.) Incubators were not particularly difficult to build, and after Tarnier published his compelling data, the idea of medical incubation for premature babies gained traction quickly throughout Europe.

By May 1897, the British medical journal the *Lancet* found the French evidence so convincing that it ran an article celebrating the imminent arrival of the first incubators to England. "The employment of incubators as a means of saving the lives of prematurely born or of very weakly infants has not yet become general in England. Yet it is notorious and obvious that the best, almost the only, means of saving such infants is to protect them absolutely from change of temperature and from cold."

A few months later, in the fall of 1897, Tarnier finally decided to retire; he was sixty-nine and had been working long hours all his life. On his last day of work, he suffered a stroke and died. He left behind several protégés, most notably Dr. Pierre Budin, who carried on Tarnier's work as care for preemies evolved and spread.

More than his mentor, Budin was very concerned with what all this progress meant for traditional motherhood, for the mother-child bond. He complained that mothers didn't spend enough time next to the incubators; he insisted that they breastfeed. He was less interested in advancing the incubator technology and more concerned with adding more glass to the existing design, so the mother could see their baby from every angle. Although he published a seminal book on medical treatment for preemies, there was an undercurrent of discomfort in his work, a sense that the technology and motherhood were at odds.

While this imagined conflict has shifted and kaleidoscoped, it persists. Incubation, from the beginning, made people a little nervous: Would the technology somehow destabilize or even replace motherhood? This feeling was captured in an illustration of an incubator show in Berlin in 1896 that is captioned "An Artificial Foster-Mother." It would be ridiculous if it weren't so compelling on some level. I felt it, too, a weird mixture of gratitude and jealousy toward Mira's incubator, the thing itself.

While Budin was in Paris trying to get mothers to sit by the incubators, a doctor in Nice named Alexandre Lion was promoting a brand-new incubator that could practically do away with both mothers and nurses, or so he claimed. Lion's criticism of Tarnier's device was that it required constant expert monitoring to maintain the hot-water bottles; with the use of a new invention, the gas thermostat, the Lion incubator could remain at a constant temperature without human intervention. The "only" care required was to feed and clean the babies. (In truth, skilled nursing was, and is, integral to any incubator's success.)

Whereas Tarnier's device looked like a box with a baby inside, Lion's incubator looked like a fancy restaurant oven. It stood tall on four metal legs, rectangular and shiny, with double glass doors in the front that showcased the baby inside. It was warmed via gas-flame-heated water, which circulated through radiator coils; a thermostat kept the interior temperature constant to within one or two degrees. A pan of water in the base kept the air humid. It had an intake that actually piped in air from outside the building, reflecting the continuing medical obsession with fresh air. The air was filtered through gauze and cotton before entering the incubation chamber and then flowing out through a chimney at the top, with a fan for circulation. It certainly looked more like a piece of technology than a cradle: Whereas Budin's design almost made the incubator disappear, Lion was making incubators that could upstage the baby.

In 1891, Lion founded a special government-funded charity nursery, the Maternité Lion, in Nice to use his expensive new incubator. There, he achieved a remarkable 72 percent survival rate. He wanted to expand, but more government funds were not forthcoming, and this care was costly.

What Lion thought of next was his really big idea: He somehow hit upon the notion of putting his incubators in storefronts, complete with babies inside, and charging the public to come and look. The admission fees would fund the care. Maybe it was the influence of the various world's fairs, which were sensationally popular and often exhibited new technology, like electricity, alongside human beings, like the racist "negro village," which was the main attraction of Paris's 1889 Exposítion Universelle.

Perhaps Lion sensed what a compelling juxtaposition this was: the shiny, futuristic incubators, and inside, the tiniest, most vulnerable people. Human ingenuity competing with human frailty, up close and personal.

Lion set up a storefront in central Paris, stocking it with incubators and babies. And people paid to see the drama. Lion opened more storefronts in Bordeaux, Marseille, Lyon. Then in 1896 he went to Berlin for the Great Industrial Exposition. There his incubator show, the Kinderbrutanstalt, or "child hatchery"(!), was a major success, drawing over 100,000 spectators, a sensation with both the public and the medical establishment.

And that was how it happened that the very next year a fellow by the name of Martin Couney got a license to use Lion's incubators in a show in London.

Couney had immigrated to New York nine years before, and it's unclear what he was doing during those first years in America or how he got the idea to go into the incubator business. One thing we do know: He was not a physician, and at first he didn't claim to be one.

Here is where we can find the thread of his story, thanks to Dawn Raffel's book *The Strange Case of Dr. Couney: How a Mysterious European Showman Saved Thousands of American Babies*: He sailed back across the Atlantic to run an incredibly popular and profitable incubator show in Earl's Court, as part of Queen Victoria's diamond jubilee. Having succeeded in London, the next year, 1898, Couney showed up in Omaha for the Trans-Mississippi Exposition, another world's fair–type event, with incubators in tow. Desperate parents brought undersized babies to him, supposedly from as far away as

Chicago. When people assumed Couney was a physician, he didn't correct them.

Then, in 1901, he surfaced at yet another elaborate fair: the Pan-American Exposition in Buffalo, New York. *Scientific American* took special note of the Buffalo incubator exhibit, describing its success and how it worked:

> *Statistics show that only about 25 per cent of the infants prematurely or weakly born live ordinarily, but by means of the baby incubator of to-day the lives of about 85 percent are saved. The baby incubator exhibited at the Pan-American Exposition . . . has proved to be of great interest to visitors . . .*
>
> *The infants are sent by the physicians of Buffalo and are given over to the care of the institution. They are weighed, clothed and placed in the incubator. They are usually under five pounds in weight on admission. The babies are taken out of the incubators every two hours to be fed by the nurses who live in the building.*

The 85 percent survival rate seems inflated, but it may be that it had a built-in selection bias, as many of the smallest, sickest babies with respiratory distress who are successfully treated today wouldn't have survived the trip. But it is remarkable to think of how compelling a treatment would have had to be to get a doctor to seriously tell a family to take their newborn down to the fairgrounds because the treatment there was the best bet.

Despite the optimistic tone of the reception in Buffalo, in other quarters there was a dark foreshadowing of the eugenic ideology that would soon grip the world. Eugenics used racist and anti-Semitic pseudoscience to falsely argue that some groups of people—like white, Christian, able-bodied northern Europeans—were innately stronger and smarter than others, and that only those with desirable traits should reproduce. It was the core belief of the Nazis. Eugenicists, unsurprisingly, were not big on incubators or preemies.

The Strange Case of Dr. Couney highlights an article in the *Buffalo Medical Journal* that was written in response to Couney's incubator show at the Pan-American Expo. "The question naturally presents itself," opines an unnamed writer, "as to whether or not this is worthwhile; whether the race as a whole does not suffer from the preservation of these weaklings to perpetuate their kind." (Count me a proud weakling who has perpetuated her kind.)

Even absent the ugly specter of eugenics, which became a powerful political force during Couney's time, this spectrum of concern—*Isn't this unnatural? Are we doing more harm than good?*—persisted and persists.

"We see more than one attitude," Baker, the pediatrician-historian, told me. "We see on the one hand some ambivalence about interfering with nature: We're going a little far with this, saving babies who are not ready for this world. And then we also see the spirit of American can-do-ism. Technological optimism." He went on: "And you see these attitudes, actually, in every culture. The technological optimism is especially strong in the United States. But so is its counterpart: the ambivalence. And you see these two strains *continue* to contest with each other." That push-pull of optimism and ambivalence was there from the very beginning.

While Couney was trotting from fair to fair, introducing incubators to America in his very flamboyant fashion, Dr. Joseph B. DeLee, a Chicago obstetrician, was trying to do so in a more conventional manner.

During DeLee's postgraduate studies in Europe, he was struck by the effectiveness of the incubator wards there. When he returned

home, he built his own thermostat-regulated incubators, modeled after the Lion version, and established his own small incubator station
at the Chicago Lying-in Hospital in 1900. After being embarrassed
by a high death rate that was mainly caused by parents bringing their
home-born babies to him too late, already half-frozen, he developed
what was probably the first transport incubator; it was basically a
briefcase insulated with hot-water bottles, enabling a doctor or nurse
to go out and pick up a newborn and bring it to the hospital without
exposing it to Chicago weather. But as DeLee refined his methods,
which included attempts to treat respiratory distress—a puzzle that
wouldn't be cracked for at least seventy more years—he found that a
high nurse-to-baby ratio was key, and that made this care extremely
expensive. He couldn't find a way to fund it: Insurance didn't yet
exist, and most medical treatment was philanthropic or paid for by
the patient. Finally, he was forced to close his clinic around 1908.

DeLee published frequently, and he ended one article about his
incubation system with a reference to several babies born quite early
who grew up to be healthy and productive, including the author
Victor Hugo. DeLee's arresting final lines are a direct rebuttal to eugenics: "Therefore, it cannot be said that the effort to save these infants is not worth making. On the contrary, every protection should
be thrown around their delicate lives."

As for Couney, when the 1901 Buffalo exposition was over, he
and his business partner were broke. There had been a disagreement
about how much rent they were to pay for their exhibit. What to do
next? He didn't want to give up the incubators: Somewhere along
the way, Couney became a true believer in a moral imperative to
treat these babies. For him, the incubator shows were not just a way
to make money or gain fame. Also somewhere along the way, he
started referring to himself as a doctor.

He wished for more legitimacy in the medical community; several sources say that he was always disappointed that his incubator shows were given space in the world's fairs' entertainment midways, with the rides, the animals, and the sideshows, instead of in the scientific pavilions. But it is not clear if Couney tried with any seriousness to establish his incubator care at a hospital or institution; it might have raised questions about his credentials he couldn't answer. But, as DeLee's saga shows, working through a hospital might not have solved the funding issue anyway.

Which is what brought Couney to Coney Island's Luna Park, an arcade and amusement complex on the boardwalk that opened in 1903. Luna Park's owner offered Couney a prime location, a good deal, and a sense of permanence, an end to the nomadic world's fair existence. Although the park closed every winter, when he simply stopped accepting babies, he could come back each spring. And so Couney opened his spotless clinic-sideshow. He would run it for the next forty years.

He installed row after row of gleaming incubators. It was staffed with professional nurses including Couney's wife, Annabelle Maye Couney, and Madame Louise Recht, a French woman who had worked at La Maternité, along with wet nurses for the provision of breast milk. He employed hawkers outside to drum up excitement; legend has it that one of them was Archibald Leach, who would soon move to Hollywood and change his name to Cary Grant.

And the babies rolled in, bedded in hatboxes and shoeboxes, clutched to chests, brought to him in frantic late-night taxi rides from hospital wards and homes. Few parents would have been able to afford this care out-of-pocket, which ran about $400 per day in today's dollars, but since the admission fees more than covered the costs, Couney provided it for free; remarkably for the time, he routinely treated babies of all races and ethnicities.

First, the babies were bathed and dressed, beribboned in blue or pink, depending on their sex. They wore tiny identification necklaces bearing their real names, to avoid accidental baby swapping, and were referred to by their initials for their privacy. Each incubator was labeled with the baby's birth weight. They slept most of the time, so still and quiet that some accused Couney of displaying dolls, or even dead babies. In a separate room, behind glass, nurses bathed, weighed, and fed their charges. Interestingly, while he emphasized strict hygiene, Couney also encouraged the nurses to hug and cuddle the babies as they cared for them, which was forward-thinking for the era.

Couney himself often provided narration for the show. He explained to audiences that, yes, these impossibly tiny babies *could* live and go on to be productive members of society. One of Couney's favorite slogans was "Maybe the future president is inside!" This is a lovely barometer for how he saw these children: full of hope and promise. He called it propaganda for preemies, and he believed it.

Couney had the greatest success with babies between two and four pounds; over four pounds and they didn't usually need much help; under two pounds and they often died of respiratory distress too quickly to get to him, or shortly thereafter. However, whenever a baby that small did show up, he did his best to treat them—it made for great PR—and with some success. It is not clear what Couney did when it became evident that a baby was dying. Raffel, the journalist who wrote a book about Couney, said that she doesn't believe babies died in front of onlookers, as part of the show. Instead, what likely happened was that a shade could be pulled down to protect a baby's last moments, or they were taken into a private area, away from view.

Some fans of the exhibit visited weekly to see how "their" babies were doing. For them, presumably, it would have been obvious when babies died: They were gone.

Madame Recht was the heart of the operation, as NICU nurses tend to be even now. She could somehow feed the smallest, sickest babies with a tiny spoon through their noses without the baby aspirating, breathing the milk into their lungs. Recht also shared Couney's flair, often slipping a ring off her finger and up a scrawny child's arm to demonstrate how small it was.

Beth Allen, now of New Jersey, was one of the smallest babies that Couney successfully treated. She was born three months early, weighing 1 pound, 10 ounces; her twin sister died shortly after birth. At first Beth's mother refused to consider Couney's care; she didn't want to leave her daughter on the Coney Island boardwalk; who *would*? But Couney showed up in person and convinced Allen's mother he might be able to help; in any case, she had no other option. Without incubation and skilled nursing, Allen would surely have died.

Allen feels that she owes her life to Couney. "I am not bothered by the fact that I was on display. At the time, that was the only way for me and the other preemies to be cared for and survive," she told me nearly eight decades later. "The paying public was necessary for the continuation of our care." Allen told me her parents came to Luna Park to visit her every day of her three-month stay, until she reached about five pounds.

Of course, thinking about Allen's story, I most identify with her mother, who was sure that her daughter would die. I imagine her handing over her surviving child to a man she had just met. (*My daughter is not a freak,* she might have thought.) I imagine her reluctantly visiting Luna Park to see her baby, surrounded by people eating ice cream, enjoying a day at the beach. What would it be like to watch them watch your baby? And then to walk away from the boardwalk, get in a cab, or shuffle up the steps to the elevated subway, leaving her there.

Despite all of Couney's success, the excitement and sense of promise that met incubators when Couney first introduced them to America started to fall away around 1910. If this story were linear, the medical establishment would have adopted incubators the moment it was proven that they worked. In fact, what happened was that hospitals realized that the nursing care involved in running an incubator ward was really very expensive. Concurrently, the rise of eugenics meant that there was no sense of urgency to help this population of vulnerable, fragile children. In fact, a "Better Babies" craze was sweeping the country; these were popular contests that graded children on pseudoscientific traits like body measurements and nose shape, crowning those that were judged "fittest." (Able-bodied children of Nordic heritage got the highest scores.) So why spend money on saving feeble newborns? It didn't help that incubators had become strongly associated with sideshows. It wasn't exactly a respectable endeavor.

Nevertheless, premature babies kept being born, so Couney's business was thriving. He took in new babies each spring and summer and shut down in the winter. (It was bad luck to be born early during the off-season.) He set up a franchise in Atlantic City, as well as Chicago's White City, where he met Dr. Julius Hess. Hess was a believer in the propaganda for preemies, too, and he and Couney become close friends.

Hess opened his own incubator ward in 1922 in Chicago's Sarah Morris Hospital for Children. Over the course of his career, which spanned thirty years, he would develop his own incubator, a warming bed that doubled as an oxygen chamber for the treatment of respiratory distress. He also wrote the first American neonatology textbook, *Premature and Congenitally Diseased Infants.* In the book's preface, Hess wrote that he was indebted to Couney. Hess is now widely regarded as the father of American neonatology, although a case could certainly be made for Couney, who never had the standing to publish his own work.

The tide changed, slowly. In a 1939 *New Yorker* profile by A. J. Liebling, Couney and his incubators are portrayed as somewhat vindicated, or at least tacitly sanctioned by the medical establishment. As a profile subject, the incubator doctor was almost too good to be true, and Liebling writes about Couney with relish: He loves a garlicky leg of lamb; he mixes a great Bacardi cocktail and is a terrible driver. "He has the firm but gentle grasp that a man might have after a life of handling canary birds." Liebling mostly allowed Couney to paint his own mythology: medical school in Germany and France, training under Budin. That biography then became accepted fact for decades.

In fact, Couney had conjured himself. Claire Prentice, a journalist who has written extensively about Couney, discovered in 2016 that there is no trace of his medical school records, dissertation, or any other paper trail that would support his claim to be a licensed physician, either in the United States or Europe. There is also no evidence of his training under Budin. Anyway, having emigrated to America at nineteen, it is extremely unlikely that he could have done any of those things beforehand.

Couney's subterfuge was convincing—and, in a way, was not subterfuge at all—because his treatment worked. Of the 8,000 babies he cared for, reportedly about 6,500 survived. His work lives still, multiplied by children and grandchildren and great-grandchildren.

The fact that the public regularly queued up to pay a not-insubstantial entrance fee just to gaze at babies behind glass is a testament to the drama of premature birth. I felt it, too, sitting by Mira's side, as she teetered on the threshold of life. There is an otherworldliness to the NICU that echoes a sideshow—something a little frightening, a spectacle you're not meant to see.

Couney was pulling back the curtain on *the* mystery: the in-between made visible. A magic show. That's why, on that long-ago Coney Island boardwalk, the tiny babies could compete with the sword swallowers and the trained lions and the screaming roller coasters.

Couney ran his clinic until 1943. It was a turning point in several ways: That same year Cornell Hospital finally opened a special care clinic for preemies, the first one in New York, although other hospitals, like Columbia-Presbyterian Babies Hospital, had by that time acquired a few incubators. In 1944, Dr. William A. Silverman started his residency at Babies Hospital. He would go on to shape neonatology as a specialty perhaps more than any other individual, making the case for evidence- and data-based practices. In 1949, standard birth certificates added birth weight and gestational age, and suddenly prematurity became a measurable problem, a public health issue.

In some ways, the history of incubation is also the history of health insurance. Charity hospitals relied on philanthropy, and they were often seen as places of last resort. Otherwise, one simply paid out-of-pocket for medical care—or put one's baby in a sideshow. Private health insurance was introduced in 1940, and it grew in popularity during the next two decades; by 1960, more than two-thirds of Americans had private health insurance. This was at least partly because there was suddenly more sophisticated and expensive medical care available, and paying out-of-pocket would have been impossible for the vast majority. Medicaid and Medicare were created by a law signed in 1965.

But even after those developments, special care for premature babies was still a dicey financial proposition. In an oral history, Dr. Louis Gluck, who founded one of the country's first NICUs at Yale–New Haven Hospital in 1960, explained why: Before 1971, insurance companies would pay for neonatal care only for babies who survived for fifteen days. (The majority of preemie deaths, then and now, happen in the first week or so.)

"If the baby survived, then they would get paid; if he didn't survive, tough. 'That's the way it is; these babies aren't worth paying for,'" Gluck told the interviewer, Dr. Lawrence Gartner. Given a financial incentive to only treat the babies they thought would definitely survive, hospitals were put in an impossible position. Then, in 1971, a bill mandated that insurance companies pay even if the babies died. And suddenly the central financial dilemma faced by every pioneer from Lion to Couney to DeLee just disappeared—although the costs were shifted, of course, onto families and society at large. "Once the insurance started paying for it, every hospital wanted one of these [a NICU]," said Gluck. "That was the big advance, no matter what anybody says. It was like wildfire once they started to pay."

Having once faced down a pile of bills that totaled nearly $1 million before insurance, I am aware that the health care system we have now is only marginally less absurd than charging admission to a sideshow. But something important *has* changed across the decades: The seesaw of ambivalence and optimism tilted. It tilted away from "These babies aren't worth paying for" and toward what DeLee wrote: "Every protection should be thrown around their delicate lives."

5

The Modern Incubator,
or How to Build a Giraffe

TODAY, THE INCUBATORS used in about 80 percent of American NICUs and in 150 other countries are called Giraffe OmniBed Carestations, made by General Electric. Each Giraffe weighs 328 pounds and costs a minimum of $10,000, although GE would not confirm an exact price. It is a clear plastic pod on a wheeled pedestal, with a monitor screen on top that displays temperature and humidity settings. The roof of the pod can be raised with the touch of a button—what some nurses call "popping the top," which sounds festive, but is often done when the baby needs a procedure. When the top opens, heaters in the lid automatically swivel on to beam warmth from above, minimizing temperature loss and turning the incubator into what is called a radiant warmer, essentially a very high-tech heated bassinet. The heaters automatically turn off when the lid is closed once more, back into convection heat, incubation mode. Via sensors, the incubator reacts to the infant's body temperature in real time, making minute adjustments.

Inside the pod is an oblong, pressure-diffusing mattress designed for fragile bodies unprepared for gravity. The bed, which is called the "Baby Susan," can rotate 360 degrees. (Get it?) It has a built-in scale and can serve as an operating table or a base for an X-ray. The entire pod itself can be raised or lowered, motoring up and down on its pedestal. It has clever ports and notches to admit the usual tubes and lines to the interior of the incubator.

The whole thing has the air of a small spaceship, or maybe a submarine, with two round portholes on both sides and one on the end, which can be pulled open to reach in to do a diaper change (an advanced move) or just to touch the baby. There is a deep-bellied storage drawer underneath the pod where nurses and parents keep supplies: diapers and wipes, gauze pads and sterile water, ointments, stick-on labels for bags of breast milk. But often both the baby and the incubator are hidden by something much more prosaic looking: special quilted covers that fit over the top of the pod to block light and sound. Mira lived in a Giraffe for two months.

It is hard for me to tease apart my first impression of Mira from my impression of the incubator itself; they seemed almost to be one organism, the way she and I had been when I was pregnant. The incubator told me how serious the situation was, but it also made me feel that Mira was being protected. I was oddly grateful to the machine itself. It was viscerally impressive—like I would imagine it would be to see a car for the first time—but also disorienting, because she was living *inside* something.

The Giraffe was an eggshell or a cocoon. When it was finally time for Mira to be in an open-topped bassinet, it felt wrong, as if she were overexposed. I missed the Giraffe's assurance of competency, of containment. The hunk of plastic wizardry that had at first seemed otherworldly had become familiar. Its logo, a friendly cartoon giraffe, is embedded in my lizard-brain memories of those days.

As I started to work on this book, I noticed that most of the other NICUs I visited used the same incubators. The cartoon giraffe was following me around. It made me wonder how this particular incubator became so ubiquitous: What's special about the Giraffe? So, one hot May day in 2017, I took the train down to Laurel, Maryland, to see the plant where the Giraffe was assembled and to meet one of the engineers who invented this incubator, about 120 years after Tarnier started tinkering with the egg warmer.

Karen Starr, a clinical director for GE and a practicing neonatal nurse practitioner, met me in the lobby of the plant, which was decorated with a giant stuffed giraffe (the animal) and large photo banners of small babies in Giraffes (the incubator). The Giraffe's name comes from the fact that giraffes are some of the quietest, most watchful animals in the world.

Starr has been working at Greater Baltimore Medical Center for over twenty years, and it's her job to help the GE engineers understand what it's like to actually work with these products when there's a real baby inside.

The way she and her GE colleagues think about NICU technology is similar to the way an airplane engineer thinks about an aircraft: Imagine a pilot at the controls. When an alarm goes off, where do her eyes go? Where do her hands rest? Can she easily reach the switch she needs while performing at an optimal level? How can the technology be designed to maximize human efficiency and safety?

"Where are the caregiver's eyes going during resuscitation?" said Starr. "That helps us understand: Do I put the monitor here if the baby's head is here? Or do I put it there?" She gestured in a different direction. "But if I put it there, every time I look at the monitor, I take my eyes off the baby."

GE asks clinicians to come in and use their products on virtual patients, like simulated flight training. They record everything, tracking the provider's eye and hand movements, and then debrief about what was efficient and what wasn't. Starr compares the resuscitation of a baby to the famous emergency landing of a plane on the Hudson River. "How many times did he [Captain Chesley Sullenberger] practice that? And he was calm. He knew what to do," she said. "The NICU feels chaotic when you are there for the first time. There are bells. There are alarms. There are people moving. But there is organization in that chaos."

Starr introduced me to Majid Akhavan, who eyed my open-toe sandals and then kindly dug up some sneakers for me, along with plastic safety goggles. Akhavan was a senior engineering manager who came from GE's aviation branch to its health care branch to oversee the manufacturing of Giraffes. He walked me down a hall to what looked like a regular office door and swung it open.

The contained office space fell away, and we were in a high-ceilinged hangar larger than a football field, a vast spotless arena with polished cement floors. And everywhere there were Giraffes in various stages of completion. Men and women were busy in assembly: screwing parts together, attaching wires to other wires, snapping plastic puzzle pieces into place. Tangles of wire in a rainbow of colors, exposed motherboards, switchboards, machine guts. Blank monitors, plastic wheels, gears, and metal bars. The floor was marked off with yellow lines delineating each worker's station; no one else was supposed to enter those spaces or risk contaminating the inner parts with a bit of sweater fluff or a loose hair. This was where Giraffes were built, in a process that takes about eighteen hours per incubator from start to finish. (In 2019, GE moved the Giraffe manufacturing plant to their headquarters in Madison, Wisconsin.)

Each worker's station had at least two monitors: One was what Akhavan called, probably for my benefit, an "Ikea monitor," which gave direction and guidance on assembly. The other was an electronic device history record (eDHR), which logged every single thing that was done to assemble or fix a particular incubator: every screw, every wire, thousands and thousands of small acts that add up to the finished object.

The eDHR can also keep track of whether or not a particular device meets the regulations of the country for which it is destined. Akhavan said that China, for instance, has recently decided that there are too many hospital alarms with red lights, which has made clinicians numb to flashing red lights. So when the new Chinese Giraffe's alarms go off, he said, the light will flash white instead. The calibration of the scale contained in the bed differs by country, too. Basically, because the force of gravity differs by latitude and by altitude, you might weigh more or less depending on where you are on earth: you weigh slightly more at the poles than you do at the equator, for instance. But a matter of a few grams up or down is important when you weigh less than two pounds, so the scales have to be minutely customized for their destinations.

Down the row, and the Giraffes started to get more recognizable. One was nearing completion but flipped upside down, showing the underside of the bed. "This is nothing but a simple heater," said Akhavan, pointing to an element that looked like a metal bow tie and a fan. "And when it flips, this is inside a bowl of water. It heats up, and steam comes out and mixes with the air." The amount of heat and humidity produced by the heater, fan, and water steamer is tightly controlled by a computer algorithm. It keeps the baby's body temperature exactly where the clinicians want it to be, within a maximum variation of only 0.5 degrees Celsius (0.9 degrees Fahrenheit).

Although this is extraordinarily precise, the actual method of heating and humidifying would be recognizable to Tarnier. I kept looking around, wishing he could see this, wondering what he would make of the space-age version of his egg warmer.

The workers produce twenty to twenty-four Giraffes per day, in shifts of either ten or twelve hours. Between 4,000 and 4,500 Giraffes are sold worldwide in a typical year. The biggest share of those sales are to the United States, but Europe, Australia, and China buy a lot and so do the affluent countries of the Middle East, like Kuwait and Qatar. "We just shipped 257 to Qatar, can you believe it?" asked Akhavan. "The country's the size of Rhode Island, but they want a lot of incubators. I hear they are building a beautiful hospital."

We were now at the far, far end of the hangar, where there were rows of fully formed Giraffes in various stages of testing and careful packing. I had a surge of weird affection at the familiar sight of them.

Steve Falk feels much the same way. He's a chief engineer and one of the people who invented the Giraffe.

In 1995, Falk was an engineer at Ohmeda Medical, a medical systems company that GE bought in 2003. He and a group of colleagues were thinking about a problem particular to NICUs. At the time, there were two different kinds of beds: incubators and radiant warmers.

Incubators provide constant warmth and humidity, but only work well when closed. If you open the top for too long—say, to put in new IV line, or even to change a diaper—the warmth escapes and the baby gets dangerously cold. To do a procedure of any length at all, you have to pick up the baby and move her to a warmer, which is an open bassinet with a powerful heater that beams down from above. The issue was that moving the babies was increasingly problematic as younger and smaller babies began to populate NICUs.

These more fragile babies weren't able to tolerate that amount of moving and handling.

So, Falk said, a handful of his colleagues got together and started talking. "Something is just weird about having an incubator and a warmer. There are reasons to put a baby in each. But they have to move between them, and we noticed that babies don't do well when you transfer them. We knew that transfer from bed to bed was a bad thing and that's all we knew."

Falk and his colleagues split up into pairs and started going around the world to watch how babies were cared for in various NICUs—big ones and small ones, at academic centers and community hospitals. They would sit and watch for hours, not saying a word, notebooks in hand. Then they'd ask questions: *Why did you do it this way? I noticed the baby struggled when this happened. What if you had to do it differently? How do the parents react to this, to that?*

On one of his first trips, he acquired a visceral understanding of the problem. "They were transferring a baby from an incubator to a warmer, and the baby passed away during the transfer," Falk said.

"The head neonatologist at the University of Tennessee told us, when you're talking about a 24-, 25-, 26-week kid: If you pick them up, they will literally try to die. They have so few calories. And you want those calories working at getting better and healthy and growing. And if they're going to have to spend their calories dealing with the environment that you just screwed up, they're going to get spent. And then they have nothing left. They desat [oxygen desaturate]. And then they crash. Literally, a metabolic chain towards death," Falk said.

By contrast, parents' touch is good for babies, when done in a developmentally sensitive way. Doing kangaroo care, for example, is usually beneficial even for the smallest preemies; their body heat is preserved by the parents' body heat, and their nervous systems are calmed by a parent's familiar smell, the sound of their voice. It's the clinical touch, the sights and smells of the hospital, the maneuvering by doctors and nurses, that can be harmful.

The clear picture that emerged from Falk's observation was that the overstimulation and temperature disruption of moving these babies from one bed to another was never ideal and in extreme cases fatal. "So a kid is in an incubator nice and toasty. You pop the top. What happens? Room air rushes in. Oh my God, this kid, his heart rate and blood pressure and sats [blood oxygen saturation] change. That's the ultimate stimulation. I just put the kid in a refrigerator. Same thing when it closes. Now room air is trapped in the incubator. How do we get the temperature and humidity back where it was? . . .

"The only way to not transfer from bed to bed is to have the baby stay in the same bed," Falk said. "I know that sounds like, duh. But the baby has to stay on the mattress. End of story."

What they needed was a bed that could convert from a closed incubator, holding in warmth and humidity, keeping out noise and germs, to an open bed that would allow providers to access the babies for procedures without losing the heat, and with as little stimulation as possible.

This is why the Giraffe was conceived: as a hybrid of the two kinds of beds. The Giraffe moves and changes depending on what the clinicians need to do, and the baby stays where she is. When Falk and his colleagues tested the prototype Giraffe in two different NICUs, babies' heart and respiratory rates, blood pressure, and oxygen saturations remained stable during the transition between the closed and open settings. "The baby, physiologically, did not feel a thing," Falk said.

The first generation came out in 2000; the second generation in 2016. The newer version is meant to be more family- and nurse-friendly: It is quieter, including hands-free silencing of alarms to minimize cross-contamination, an easier-to-read monitor, and decreased fan noise. But Falk says that the algorithm that directs the transition between incubator and radiant warmer remains the unique, defining feature of the Giraffe, the thing that makes it so popular with clinicians.

It's easy to look at a piece of technology and see it as a cold, inhuman machine, when in fact it is nothing but a manifestation of human experience. In those first days after my C-section, when the nurse lowered the Giraffe down to the level of my wheelchair so that I could see Mira, I didn't know that feature existed because a mother in a focus group told Falk how she couldn't lay eyes on her baby for four days because she couldn't stand up and the incubator couldn't be lowered. I didn't know the difference when they popped the top and the heater on the lid automatically whirred to life and started beaming heat onto her. I didn't know about the baby who died as Falk watched, who left him with such a vivid impression of the problem he was trying to solve. But all these stories were embedded in the machine, invisible.

6

The Incubators
of the Future

Babies in Bags

IN 2017, RESEARCHERS at the Children's Hospital of Philadelphia (CHOP) published a study in the journal *Nature* on an extrauterine system they called the biobag, but which everyone else called an artificial womb. Accompanying the paper was a video.

It shows a lamb in a bag. There are few tubes or lines visible, save for a thick catheter into the lamb's belly, right where the umbilical cord would be. The animal is sealed in a clear, fluid-filled rectangular plastic sack, roughly the shape of a very large ziplock bag. One knobby knee is crooked; the eyes are shut tight. The lamb appears to be skinned because it is raw pink all over, but in reality it is just immature, a fetal lamb that has not yet grown wool. When I first saw it, the only frame of reference my brain could conjure up for what I was looking at was cooking: The lamb looks like it is about to be braised; it seems destined for a sous vide cooker. But then it moves. It kicks its tiny hooves and the plastic around it ripples. It twitches like a dog dreaming. Its muzzle nudges something invisible and it appears to chew and swallow as its little ears flex backward. It is here and yet not here, born and yet not. Premature.

The lamb-in-a-bag is a deeply strange sight. It evokes a long line of science fiction clichés, from the child hatchery of *Brave New World* to the baby pods of *The Matrix*. As symbols, artificial wombs represent a fundamental severing, an imagined future in which technology has altered the most basic definition of what it means to be a human. When we look at the lamb-in-a-bag, we see much more than just a lamb in a bag.

BUT DR. KEVIN Dysart, a neonatologist and one of the researchers on the project, said that the main idea behind the biobag is grounded in the natural: "Let's just keep you the way you were."

Currently, neonatal life support treats premature infants as though they are newborns who need help doing everything that newborns should be able to do. So we blow oxygen into lungs not ready to breathe, we tube feed a digestive system that's not ready to digest, we keep immature bodies warm and hydrated using heated incubators. The biobag treats extremely premature babies—or, for the time being, extremely premature lambs—as what they *should* be: fetuses. Fetuses are designed for a certain environment, and the biobag tries to replicate it.

When I sat down with Dysart—a gregarious Philadelphia native—he tried hard to steer the conversation away from science fiction, a motherless world. He doesn't see the biobag as a symbol or the first step down a road to dystopia—or utopia, for that matter. He sees it as a medical treatment for a very specific population. He thinks it has the potential to be a game changer for extremely premature babies, maybe even more so than all the innovations that have come before. He and his colleagues object to the term "artificial womb" because they don't want to suggest that the device can replace mothers. Instead, they believe the biobag is a treatment that can bridge a gap for babies born on the most dangerous threshold, between 22 and 25 weeks.

A fetus in the womb relies on the placenta for both nutrition and gas exchange—carbon dioxide out, oxygen in, the function that our lungs take over when we are born. The fetus is tethered to the placenta through the umbilical cord, and umbilical blood carries waste products out and brings oxygen and nutrition back to the fetus's body. Meanwhile, the cloistered, liquid environment of the amniotic sac protects the fetus's permeable skin. It allows the entire body—crucially, the lungs and brain—to develop normally, and the mother's body temperature keeps the fetus warm.

It is a natural impulse to want to put premature babies back into warm water; in fact, the first patent for an artificial womb was granted in 1955. (It was apparently never used.) But it is a very difficult thing to accomplish technically. For one thing, once a baby is exposed to air and tries to breathe, their respiratory and circulation systems change, and they can't go back to being aquatic creatures who don't use their lungs. In order to keep them as they are—to put them in the biobag—the baby's body has to essentially not realize it has been born.

Artificial placentas and fluid environments have been tried before in premature goats and lambs, with very limited success. (Sheep and goats might seem like unlikely harbingers of the science-fictional future, but they are often used in neonatal research because their respiratory and circulation systems are similar to those of human babies.) The CHOP group seems to be the first to claim that a version of their model will soon be ready to use with real babies inside.

(In 2017, CHOP lead researcher Dr. Alan Flake said that they could be looking at a clinical trial in one to two years, though that has not yet come to pass.)

Previously, the most notable attempts came from researchers at Juntendo University in Tokyo and at the University of Tokyo, who kept premature goats alive in various artificial womb systems starting in the 1980s. A paper published by Dr. Yoshinori Kuwabara in 1987 described fourteen premature goat fetuses delivered by C-section and kept alive in warm, fluid-filled tanks, with catheters threaded through their umbilical stumps and into an artificial placenta-like membrane that oxygenated the blood and ran it back into their bodies. The goats were kept alive and in good condition like this for up to about six days before their hearts failed.

Other Japanese researchers built on Kuwabara's work. (One paper from 1998 attempts to strike a reassuring tone: "The mother goats were returned to the farm after recovering from the operation.") These groups had some success, but ultimately had problems with circulation, or with normal development—some baby goats ended up deformed. In some cases, the system required that the goats be sedated so they didn't move, which interfered with normal maturation. Gestation is almost infinitely complex, and even the best scientists in the world struggle with its intricacies.

But the group of eighteen scientists at CHOP, led by Flake and Dr. Emily Partridge, felt that there was still a possibility this kind of model might work. They started from scratch, tinkering with tubes and tubs they bought from a beer brewing supply store, combined with parts scavenged from an extracorporeal membrane oxygenation (ECMO) machine, an apparatus that can be used to temporarily bypass a person's heart and lungs if those systems are failing. ECMO machines pull blood out of a patient's body, pass the blood through a membrane that removes carbon dioxide and adds oxygen, and then return the oxygenated blood to the body. It has an obvious parallel to the placenta.

Many previous attempts to do this, though, relied on a pump to circulate the blood out of the fetal lamb's body through the oxygenating membrane and then back to the body. But the pump messed with the delicate fetal blood pressure, causing high rates of hemorrhage and cardiac problems. Other attempts to try a pumpless system resulted in a failure of the circuit itself; there wasn't *enough* pressure to keep the blood moving at the right rate.

This time, when the CHOP team rigged up the first incarnation of the biobag, they found that they could just use the fetal lamb's blood pressure to pump the blood out and back. Newer oxygenating membranes had less resistance than the older ones, so it was easier for blood to pass through. The lamb's heart was the only pump necessary to maintain the circuit. They also invented a new kind of umbilical cannula that allowed the lambs to move without the worry that they would become unhooked from the oxygenator.

Creating homemade amniotic fluid presented its own challenge; eventually the researchers came up with an electrolyte solution—water mixed with elements such as sodium, chlorine, and potassium. But because the fluid is essentially warm water, bacteria loved it. After a few lambs died from massive infections, the researchers realized that the fluid would need to be continuously exchanged—flowing out and being replaced—to keep it sterile. That means the system requires a lot of homemade amniotic fluid: the researchers had to make more than three hundred gallons of it a day.

In a five-year process of trial and error, the CHOP researchers finally succeeded in gestating fetal lambs in the bags for up to four weeks. The lambs moved; they swallowed; they practiced breathing as all fetuses do; they grew wool and opened their eyes for the first time. They did all this in the bag, as they would have inside their mothers. When the researchers "delivered" them, they found that most had normal body, lung, and brain growth. Although they euthanized most of the lambs to study their systems, at least one lived for a year, with continued normal brain development.

Lambs gestate for about 145 days, and the lambs in this study were placed in the biobags at between 105 and 117 days' gestation, the respiratory developmental equivalents of a fetus at about 23 to 24 weeks. The ideal human patient would be an extremely premature baby, one of the babies born in the viability zone: 22 to 25 weeks' gestation.

No one thinks the bag would be better than a real uterus, but they do think it might be better than an incubator and a ventilator. The purpose would be to carry the tiniest babies through the worst of the immaturity, when mortality is highest and morbidity the most severe, and then transfer them to a traditional incubator when they are about 28 weeks, the turning point at which prognoses are typically no longer dire.

The researchers imagine it working like this: An extremely premature baby would be delivered through a modified C-section, similar to the ex utero intrapartum treatment (EXIT) procedure that is now used in fetal surgery. This makes it possible to partially deliver and work on the baby while the child is still connected to the functioning placenta. It would allow doctors to insert the cannula into the umbilical vessels before putting the baby directly into the bag, so the baby never takes a breath. The process would take less than two minutes. At that point, the baby in the bag would be put into, essentially, a clear box and placed in a dark room. Physical exams would be done via ultrasound (just like they are during pregnancy) once or twice a day. A night vision camera would allow parents to see their baby in real time, and the sounds of the mother's heartbeat could be piped in.

Among scientists who are not involved in the research, the feeling seems to be a cautiously excited *Wait and see.* "So Flake has his lambs, and they look very cute scampering about in the fields after they come out of their bags," Dr. John Lantos, a prominent neonatal bioethicist said, half joking. He went on more seriously: "Now to find out whether it'll work in people or not. It's hard to say."

Dr. Steven McElroy, a neonatologist at the University of Iowa, echoed the sentiment: "It's really far from prime time. I think it shows a lot of promise; it's really interesting. But they've got to go a long way from doing it on lambs to doing it on a 23-weeker. That may be the next big thing. It may not. But I think there's excitement about the potential."

Karen Starr, the neonatal nurse practitioner who advises GE, said that she and her colleagues had talked with CHOP about the biobag. (If the tool actually works for humans, it is reasonable that eventually the CHOP researchers, who have the patent, will have to license a company to manufacture and sell it.)

Starr found the research fascinating and thinks that this kind of technology will be used with premature babies in the future, though probably not within the next five years. She, too, was waiting for more evidence. And, from the practical end, she wondered: How would a biobag change workflow in a NICU? How would nursing staff react? What would the space requirements be? What would parents think?

The general public seems to find the idea of the biobag much wilder than the clinicians do. This is at least partly because there's a lack of understanding and context for what the biobag is trying to accomplish. Not many people have encountered a 22- to 25-week baby. Not everyone knows that NICUs are already fairly wild, futuristic places. Not everyone has wondered if their baby was too immature to live in this dry, gravity-heavy world. Then again, the biobag really *is* pretty wild.

Judging from the press coverage—and there was a lot of it—there was, on the one hand, amazement and, on the other hand, a barely suppressed panic about an inexorable first step into an unrecognizable future. What did this mean for the viability line (and, thus, abortion law)? Would mothers become obsolete? Could this bag be used to grow humans from start to finish? Would this mean women's liberation or oppression?

According to the creators, the biobag wouldn't change *anything* except to improve the lives of babies born between 22 and 25 weeks.

"All right, a baby is going to be born really early," said Dysart. "We need to exchange gas for them. What would you rather do? Make it like a placental environment or bang away at them with a ventilator? You wouldn't *choose* bang away with a ventilator, but we've been doing it for decades, so we keep doing it. And when we change, it's rough. Change is hard."

Lambs are not people, but still: It is very true that the lamb dreaming in its bag looked a lot more comfortable and relaxed than any ventilated premature baby I have ever seen.

Every new treatment for premature babies brings up the question of viability and whether or not it's going to shift earlier. The researchers insist that their invention can't be used on babies younger than 22 or 23 weeks. Flake acknowledged that any new treatment brings up new ethical questions but said he would be very worried about using the biobag at lower gestations. First of all, he said, it would likely not work, because at lower gestations fetuses' blood pressure and flow wouldn't be sufficient to pump the circuit. It would be more likely to inflict harm than help them survive.

Dysart was more direct: "I mean, they don't have skin," he said, speaking of fetuses younger than 22 weeks and the impossibility of treating them. "You'd have to have a whole other tool."

None of this explanation stopped the speculation about the big picture, the long term. It is true that the whole point of medical research is that scientists build on the work that has come before. So CHOP researchers may not intend their invention to be used on a 20-week—or 18-week, or 17-week—baby, and they may not think it would even work with babies that immature; but once a technology is out in the world, its creators don't have control over it. It evolves. And if the biobag works as they hope it will, it would mean that 22- and 23-week babies might have outcomes comparable to 28-week babies. If—*if*—that hope pans out, it's easy to imagine that parents of 21- and 20-week babies would ask to put their children in the bag, too.

In some cases, the news of the biobag seemed to serve as a starting point for a referendum on how people feel about premature babies and whether they *ought* to survive: An article in *Maclean's*, a Canadian news magazine, covered the news of the biobag not by meaningfully evaluating the biobag itself but by wondering incredulously whether the lives of extremely premature babies are worth saving. The author cited as evidence that 90 percent of surviving 25-week babies are severely impaired—a statistic that is so wildly inaccurate that to me it suggests bad faith. (The proportion of survivors at that gestation who are *moderately* to severely disabled is more like 39 percent.)

It also sparked questions about the mother-child bond. A 2018 story in the *Irish Times* asked, "Might AWT [artificial womb technology] further challenge the apparently inviolable bond between mother and baby?"

But what *is* the inviolable bond between mother and baby? Is it utterly dependent on 40 weeks in your very own uterus? Not all parents gestate their children. If we use medical technology to keep babies alive so they can go home with their parents, is that really a violation of the mother-child bond? If the longed-for baby dies, surely *that* violates the bond? What is the difference, in the end, between a fluid-filled bag and an incubator filled with warm, humid air?

Elizabeth Chloe Romanis of the Centre for Social Ethics and Policy at the University of Manchester in the U.K. thinks there is a major difference. In an article in the *Journal of Medical Ethics* she took issue with what she saw as the CHOP researchers attempting to deny a fundamental ethical divergence between the biobag and current NICU care. She thinks a child in the biobag might be a new kind of patient altogether, neither fetus nor baby, and proposed the term "gestateling." This is the crux of the difference as she sees it: "Instead of assisting a premature neonate with functions it is struggling to perform alone, AWT [artificial womb technology] treats its subject *as if it had not been born*."

If the patient has not really been born but is no longer a part of the mother's body, then what rights does it have and who gets to say? Most legal and ethical traditions treat fetuses and babies differently, with different rights.

But these questions are variations on the dilemmas that neonatology faces already. In some ways, for better or worse, we already treat extremely premature babies differently from other children: They are already occupying a nebulous space between fetus and baby. I am not sure it matters what we call the occupants of the bags. Each one of them will still ask us the same questions that extremely premature babies ask us now: What is best for them? Who gets to say?

The CHOP researchers referred to the lambs-in-bags as "extreme premature fetal lambs" but called their future target population "infants." They see the biobag fitting into the current ethical framework of the NICU, where the patients are babies, with all the rights that people have—albeit people who can't make their own decisions and must have proxies, usually their parents, make decisions for them. This system works, although it has its own complications.

There is nothing new about the reaction to the biobag. The invention of the incubator was met with many of the same concerns. There have always been those who argued that the weaklings are not really meant to live; that they will be disabled, a drain on society; or that incubation is unnatural, that it will erase motherhood. The biobag evokes the exact same reaction that Martin Couney had to battle—*This isn't natural*—as well as the same reaction that he profited from—*This is incredible*.

In Dr. Jeffrey Baker's book, *The Machine in the Nursery*, he writes that "as the incubator became more complex mechanically, it acquired a threatening aspect. The metaphors of the artificial uterus, nurse, and environment symbolized the effort of science to imitate and improve upon nature." Here he is referring to a fairly primitive early incubator, but it could just as easily apply to the biobag.

The fear of this imagined threat can be so pointed that it can derail inquiry altogether. Dr. Helen Hung-Ching Liu, a reproductive endocrinologist, worked on this problem from the other end: infertility. An in vitro fertilization pioneer at Cornell, her research was mainly concerned with how embryos implant in the uterine lining, and the reasons that this crucial first step in pregnancy can go wrong. She grew endometrial cells in her lab—layers of them—studying the interaction between uterine lining and embryo. In the end, she had made a kind of artificial womb.

In 2001, Liu successfully gestated a human embryo for ten days in her scaffolding of endometrial cells, but in the United States it is illegal to experiment on a human embryo that is more than fourteen days old, so that was the end of that. But in 2003 she grew a mouse all the way from fertilization to near term in the artificial womb. The mouse moved around and practiced breathing, as all fetuses do. Maybe it dreamed baby mouse dreams. As far as I can tell, it was the first mammal to lack a gestational mother of any sort. But the creature was also deformed—a 2005 *Popular Science* story that details Liu's experiment says it looked more like a seahorse than a rodent—and it quickly died. The same thing happened each time Liu repeated the experiment.

You might wonder why you haven't heard about Liu and her motherless mice. (Talk about a sideshow.) It's because, according to press accounts, she declined to publish her results, under pressure from many sides.

I couldn't find Liu. For many years she was the director of the Reproductive Endocrine Laboratory at Weill Cornell medical school, but when I called there, the center had no record of her. Her LinkedIn profile still lists her as a professor at Weill Cornell Medicine, but she is not listed in the school's directory. Her Cornell email address bounced back, and when I emailed Cornell public affairs asking about her, they emailed me back a stone wall. ("We are unable to participate in this request at this time.") When I tried to contact Liu's former colleagues and coauthors, they either didn't respond or passed me off to public affairs, who sent the same email again, word for word. I called past numbers listed as hers to find they had been disconnected. I asked journalists who had written about her in the past, but they had only her Cornell email address, which wasn't working any longer.

Because I couldn't contact Liu, and because she didn't publish her most stunning research, I can only wonder at the scraps of her story she left behind. Liu got a lot of media attention over the years she was working with artificial wombs. Every time she was quoted, it was with optimism, with certainty that, with this new tool, she would be able to help women who can't conceive, perhaps by transplanting the lab-grown uterine tissue along with the implanted embryo into a woman's body. (She seemed to have come to think that growing babies entirely outside a mother from start to finish would not work.) She was quoted multiple times talking about her patients who were desperate with longing for a child—how she got phone calls from women asking to volunteer for her research, willing to do anything for a baby.

The last place I can find Liu is in a 2015 article in the *Atlantic* by Katherine Don. At that time Liu was still working at Cornell. Don wrote that, despite Liu's success, she had halted her research and refocused on other topics. "There was a lot of pressure from the press," Liu explained to Don. "Everyone was talking about it. The medical ethicists were against it. Pro-life people were against it, and pro-choice people too—both sides. This came as a surprise to me. When I started, I just wanted to help women who had implantation problems. But it turned out to have all of these social implications, and I didn't want to deal with it."

With every new gestational technology, some people see a slippery slope where others see progress. And some people, understandably, can't afford to see it in societal terms: someone who is hoping against hope after a miscarriage, for instance, or after a stillbirth, or after having an extremely premature baby who died.

The Philadelphia researchers who invented the biobag said that they were in talks with the FDA to start human trials within the next year or so. As of this writing, there is no FDA record of a human trial having started, which would be public information. But I have been unable to get an update from the researchers themselves. After our first conversation, the researchers abruptly stopped wanting to talk.

When scientists come up with new ways of treating, say, cancer, the conversations don't seem to be so fraught. But treating premature babies brings up unique, fundamental questions about existence. The lamb-in-a-bag will never be just a lamb in a bag. I admit, though, amid all that complexity, what I see most in the biobag is an attempt to prevent suffering—a wish that the smallest babies (or fetuses, or gestatelings) can keep dreaming their unknowable liquid dreams.

Part III

The Breath

Treating Respiratory Distress

Dr. Mildred Stahlman and
the Miniature Iron Lung

THE TINY IRON lung languished in the Vanderbilt University Medical Center basement for five years until October 31, 1961, when Martha Humphreys was born. She weighed about five pounds, a very good size, considering that she was eight weeks early—meant to be a Christmas baby, born a Halloween baby. But almost immediately she started to gasp and struggle for breath. Everyone knew what that meant, including Martha's father, Jerry Humphreys, who was a fourth-year med student at the hospital, and Dr. Mildred Stahlman, the pediatrician who ran the nursery.

It meant that Martha didn't have much of a chance. Premature babies who gasped for breath had what was at the time called hyaline membrane disease and is now called respiratory distress syndrome (RDS). Most babies like Martha, who had a bad case of it, died: They turned gray-blue and panted and heaved for hours or days before they became exhausted and simply stopped. There was no effective treatment. Sometimes an oxygen mask or chamber helped the babies get through the worst of it, as did placing a bag over the nose and mouth and squeezing as a kind of artificial mouth-to-mouth, but often those interventions only prolonged the process.

Stahlman, a slight, intense woman with sky-blue eyes who stood under five feet tall, was one of the only doctors in the world at the time who knew how to thread a catheter into an umbilical vessel, allowing her to precisely measure the oxygen saturation of a baby's blood. (She had studied pediatric cardiology in Sweden, a country that was and is on the forefront of neonatal medicine. There she acquired both rare knowledge and liberal political tendencies.) She had been researching fetal and premature lambs; she kept the pregnant ewes out in the hospital courtyard. She had been working doggedly toward this moment when she might have something to offer a baby who couldn't breathe.

The iron lung, or negative pressure respirator, or tank ventilator, works like a bellows to mechanically pull lungs open through intermittent negative pressure. It's actually how our lungs naturally work: The muscles in our torso expand, creating a vacuum in our lungs that then easily fills with the air we breathe in. To use a negative pressure respirator, you seal a person's entire torso inside the machine and cycle the pressure (vacuum, then release; vacuum, then release) to pull their lungs open and closed. It was famously able to keep polio patients alive when paralysis prevented them from breathing on their own.

After the eradication of that disease in the United States around 1955, the Monaghan Company, which made the tank respirators, had a bit of a marketing problem. Who else needed help breathing? Preemies. So the company miniaturized their machine, made four prototypes, and sent them around to doctors who specialized in babies, including one to Stahlman at Vanderbilt, where it languished in that storage room until Martha's birth.

No one can say why Stahlman chose exactly this day to pull out the iron lung. The timing was good in that she had opened a small lab next to the nursery, thanks to a grant from the National Institutes of Health. There she was deepening her study of respiratory distress and how it manifests itself in babies' blood gases. Maybe it was because Martha was relatively big; she would be more resilient than a smaller baby, more likely to do well. Maybe it was because Stahlman knew that what she was about to do was basically an experiment, and Martha's father understood this, too, and still wanted to try it. Maybe it was finally one baby too many.

If you could miniaturize yourself and slide down a trachea, down through one bronchial tube into a lung (left or right, take your pick), and keep going as the branches divide and get smaller and smaller and smaller, you would arrive in the end in a microscopic air sac—one of about 400 million such sacs in each adult lung, or half that number in healthy term newborns. These sacs are called alveoli. This is where the air from outside arrives, laden with oxygen. And as this sac expands with air, oxygen diffuses into the blood. On exhalation, the sac deflates to push out carbon dioxide.

If you could stand inside an alveolus, you'd probably be slipping and sliding around as though you were inside a balloon in which the interior had been rubbed all over with liquid soap. That's because, in healthy, term infants (and healthy adults), the sacs are lubricated with a viscous, soap-like substance called surfactant. When the sacs deflate, they don't collapse in on themselves and don't stick together; instead, they have the slippery flexibility of a soap bubble, so they easily slip open and closed, open and closed. They are elastic, inflating and deflating smoothly and without effort.

Gas exchange: oxygen in, carbon dioxide out. In the womb, a fetus's lungs are filled with amniotic fluid. The placenta handles that critical in-out through the umbilical cord. Once the baby is born, however, they must undertake gas exchange for themselves, using their lungs. Some of them are not ready.

Premature babies' lungs haven't finished maturing. One of the biggest problems with this is that the surfactant hasn't developed yet. This is quite common in babies born before 30 weeks, but it can happen with more mature preemies, too, like Martha.

Now imagine you are inside a balloon, but this time it is much less elastic and there's no soapy substance inside. When it deflates, the rubber collapses and sticks to itself. It is stiff and hard to inflate, rather than supple and slippery.

In a preterm infant, when the air sacs collapse on each exhalation, the alveoli fold in on themselves brittlely and become more and more difficult to reinflate. These babies have to work so hard to breathe that their rib cages stand out sharply with each inhalation, like gills. They pant, breathing in excess of 100 times a minute, and turn grayish-blue. In the absence of treatment, many or most will die within three days, depending on the severity. Their lungs collapse and they can't get them open again. If, on the other hand, you can find a way to help them breathe for three or four days, there is a good chance that their lungs will start to mature on their own, and they may survive.

Breathing is often the most acute problem preterm babies have. There's no time to think of brain development or nutrition or anything else if the baby can't breathe. The fact that the smallest babies, the ones born early, have breathing problems must have been evident to anyone caring for them in any era, because their distress is so painfully easy to see.

The difficulty was clear to Mabel Bell, Alexander Graham Bell's wife, who in 1881 gave birth to a premature baby boy, Edward, who struggled for breath and lived only three hours. Mabel was devastated and wrote in her diary, "He was a strong little fellow and might have pulled through if they had once established regular breathing."

She already had two little girls but longed for another baby. In 1883 she gave birth in her seventh month of pregnancy to another boy, Robert. "Poor little one," she wrote. "He was so pretty and struggled so hard to live, opened his eyes once or twice to the world and then passed away." Both parents grieved over these lost babies for the rest of their lives. In *Reluctant Genius*, a biography of Graham Bell, author Charlotte Gray cites Mabel's diary and the letters written between husband and wife, in which they are both consumed with sadness. "Nothing will ever comfort me for the loss of these two babes," Alexander wrote to Mabel in 1885.

Graham Bell is known for inventing the telephone, but he was a prolific inventor of other devices as well. Shortly after his first son's death, Graham Bell developed an intense interest in building an apparatus for artificial respiration.

He devised what he called a vacuum jacket, an airtight cylinder that could be fitted over a baby's torso and connected to a bellows that could be pumped to expand and then contract the baby's rib cage. He made several versions of this jacket over the years and tested them on small, intentionally half-drowned animals, like sheep, with some success in reviving them.

Graham Bell explained his invention like this: "Many children, especially those prematurely born, die from inability to expand their lungs sufficiently when they take their first breath. I have no doubt that in many of these cases, lives could be saved by starting the respiration artificially by means of an apparatus operating in the manner described above." His assessment of the physiological problem was not quite right but close enough. His vacuum jacket worked by the same negative-pressure mechanism as the small iron lung.

Around 1889, Graham Bell presented the jacket to the American Association for the Advancement of Science, but no one except him seemed particularly interested in solving respiratory distress in babies. Nevertheless, Graham Bell continued to sketch and build prototypes of tiny respirators for years. His inventions were apparently never used on actual babies, and the Bells never had another child.

And so it went, for a very long time. Hyaline membrane disease (now RDS) was the leading cause of death in preterm babies throughout the first two-thirds of the twentieth century, killing about 25,000 babies a year.

"Hyaline" comes from the Greek *hyalos*, or glass. For decades doctors thought that the problem was caused by the presence of something extra, a glassy, brittle substance in the lungs of tiny babies who had died. On autopsy, it was clearly visible: The lungs were collapsed, airless. And inside them, in the spaces that should have held air, were hard membranes plastered to every surface.

Starting in the late 1950s, Dr. Mary Ellen Avery hypothesized that, in fact, it was the absence of surfactant—not the presence of something extra—that was causing the lungs to collapse and crystallize. But knowing what the problem was didn't solve it.

It was not for lack of trying. As soon as incubators advanced enough to double as oxygen chambers, like the Hess bed that was first used with oxygen in Chicago in 1931, clinicians filled the closed beds with the piped-in gas and immediately saw improvement in mortality rates; at Hess's premature infant station, survival of preemies overall went from 52 to 79 percent when oxygen was used. Some doctors strapped funnels to tiny faces and delivered oxygen that way. At least one doctor even started injecting babies with oxygen, although that was short-lived and not effective.

Giving oxygen like this to treat RDS is a half measure. It can mitigate respiratory distress and apnea, but in RDS's severe form, when a baby's lungs are collapsed and unable to inflate, it doesn't matter how much oxygen is available in the environment, because the baby is unable to breathe it in. But it was all doctors had to offer for decades.

Neonatology was a brand-new discipline—it didn't even have a name yet—and it was basically the sum total of the whims, experience, and judgment of several dozen pediatricians who were providing the only treatments available, at the few hospitals that had special incubator nurseries. There were no randomized controlled trials; there was no parental consent or process of shared decision-making. The systems that we take for granted now simply didn't exist; the relationship between patients (or parents) and doctors was more patriarchal. Doctors were the authority figures, and they decided how things would go. And anyway, parents didn't have much choice: bring the baby home, probably to die, or trust the doctor to do things like inject your infant with oxygen.

In the early 1940s a new idea emerged among this small band of doctors: They were routinely giving babies gas containing 40 percent oxygen. (The air we breathe is 21 percent oxygen.) If 40 percent oxygen was helping with respiratory distress, wouldn't 70 percent be even better? And why restrict the treatment to babies that were turning blue and gasping? Even preterm babies who seemed to be breathing comfortably had been shown to have lower oxygen saturations in their blood than term babies and adults. So the newly established premature baby incubator units started bathing babies in extremely high concentrations of oxygen as a routine treatment.

In 1941, Dr. Stewart Clifford of the Boston Lying-in Hospital was following up with two former preterm babies at several months of age; both of them had mysteriously gone blind. They had the same distinctive symptoms: a buildup of scar tissue on the surfaces of their retinas, which eventually detached, giving their eyes a cloudy, white appearance. A third case quickly popped up. And then another and another: more cloudy-eyed survivors—all of them in the cities where preterm babies were treated in the advanced new incubator wards, such as Boston, New York, and Baltimore. What followed in the next twelve years was the rampant spread of this disease, called retrolental fibroplasia, and an increasingly frantic search for the cause.

Over the course of those years, 10,000 children were blinded (Stevie Wonder among them) before several studies suggested that exposure to high concentrations of oxygen damages the blood vessels in the retina, causing scarring and blindness. This is now called retinopathy of prematurity (ROP), and strict oxygen protocols have dramatically decreased its prevalence. However, about five hundred premature babies are still blinded by ROP each year in the United States; not all the causes of the disease are fully understood even now.

The saga of ROP is extreme but emblematic of the challenges of neonatology, which is why the pioneering neonatologist Dr. William A. Silverman wrote an entire book about it, *Retrolental Fibroplasia: A Modern Parable.* Every treatment exacts a cost on the baby's body. Even now, the exact concentration of oxygen to give to babies in severe respiratory distress who are not responding well to standard treatment is controversial. And blindness is preferable to death. After the ROP outbreak, neonatology was chastened, and back to square one: modest oxygen concentrations, given through a mask, a bag, or an incubator. And there was an increasing sense, led by Silverman, that if this new field were to be organized into a specialty, it would need to be more systematic and evidence-based— relying on research rather than gut feeling.

There were other attempts to find new solutions, some of them gruesome. A short-lived and horrifying treatment from the 1950s involved trying to physically hold a baby's lungs open by looping a suture through the breastbone and then clipping the suture to the top of the incubator, trying to pull the chest open by traction, or suspension. This was done without pain medication. In 1964 one doctor claimed that Epsom salt enemas were a miracle cure for RDS. It was reported as fact in the *New York Times* before closer scrutiny found that it was not only ineffective but also likely to cause death.

By the early 1960s, a handful doctors in the U.K., Sweden, and South Africa had attempted mechanical ventilation—rudimentary life-support machines that could, one way or another, breathe for the babies. They had limited success, a few survivors. Word was slowly percolating out, but there was no collective sense, certainly no certainty, that this was going to be *the thing* that worked. For all anyone knew, it might be another case of chest traction or oxygen injections. In North America in particular, the idea was still considered extremely experimental.

Stahlman, in Nashville, with her little iron lung in the storage closet, was ready to step into that breach. She knew that they had to work quickly to save Martha and to prevent brain damage from lack of oxygen. So she explained to Jerry Humphreys that his daughter would die without treatment but that she wanted to try the iron lung. Her research gave her good reason to think it might work, but she had never done it before, so she really didn't know what would happen. Jerry agreed, so in went Martha: her body into the vacuum chamber, the seal around her neck, leaving her little head sticking out. Stahlman didn't go home that night, or the night after that. She sat by tiny Martha and the machine.

Here's how Stahlman herself remembered Martha's first few days when she wrote about it almost two decades later, in 1980:

Within a few weeks of opening our laboratory, an infant was born in our hospital with severe hyaline membrane disease. It happened to be the baby of a senior medical student. We thought the infant would surely die, despite oxygen by mask and scalp-vein buffer, which were the extent of our therapeutic armamentarium at the time. We offered to put the baby in our negative-pressure tank-respirator, if the parents were willing. They agreed, and after placing an um-bilical venous catheter in the left atrium (we were unable to enter an artery), the baby was sealed in the tank where she remained for five very tiring, but enormously instructive days. She was success-fully weaned from her ventilatory assistance, and much to our relief, her umbilical catheter could be removed, and she is now a beautiful 20-year-old girl with an I.Q. of 140.

The iron lung worked, and Stahlman kept going. By 1965 she was able to present data showing that she had put twenty-six actively dying babies into the tank respirator for a period of time ranging from a few hours to ten days, and eleven of them survived. (With-out the treatment, it is quite likely that they all would have died.) She found that most of the survivors needed three to five days in the respirator and that she could wean them off the assistance by turning the vacuum off and letting them breathe on their own until they got tired, then turning it back on again, repeating this process until it seemed that they could go it alone.

Meanwhile, by virtue of using this technology, Stahlman was cre-ating one of the first NICUs. What really makes an intensive care unit an intensive care unit is not how sick the patients are but the amount of technology being used to care for them and the level of skill necessary to run the technology. The tank respirators—and the positive pressure ventilators that Stahlman would eventually also employ—were not easy to use: The babies' blood oxygen and carbon dioxide levels were not easy to monitor. That's why Stahlman herself almost never went home, never slept, in those early years. An extremely skilled clinician had to monitor the babies around the clock. It wasn't like a simple incubator ward, in which the infants would mostly sleep and there wasn't a lot that anyone *could* do.

If Stahlman wanted this new kind of intensive care for babies to be sustainable and replicable, she had to train other clinicians to do this work. Eventually, she would become known for her dedication to her fellows, more than eighty pediatricians whom she trained to specialize in neonatology over the years.

Dr. Håkan Sundell was one of them. In 1963, just as Stahlman was having more success with her tiny iron lungs, he graduated from medical school in Sweden and started working at a hospital in Stockholm, where they didn't have any such treatment. He would lie awake at night and worry about the littlest babies who couldn't breathe.

"Oh, it was very hard, very hard. It was so dramatic," he said. "The babies were obviously suffering very much; they're working so hard to get enough oxygen. Eventually they just gave up. They were too tired and stopped breathing."

When Sundell heard about Stahlman's work treating RDS and managing the babies' delicate respiratory and circulatory systems, he was captivated. It was amazing, he said, exciting. In 1966 he went to Vanderbilt to train as a neonatology fellow under Stahlman. He went on to become an attending neonatologist and professor of pediatrics at Vanderbilt, working closely with Stahlman for decades.

"She's a very intelligent and brave woman. She has extraordinary intellect," he said.

Stahlman, who is now in her late nineties, remains a bona fide legend in Nashville, the iconoclastic, slightly terrifying daughter of an old and wealthy newspaper family who defied social expectations and chose not to marry in order to dedicate herself to premature babies. Her colleagues call her Millie, a diminutive rather at odds with her tough, uncompromising reputation.

I went to Nashville on a sunny, unseasonably cold October day to visit the NICU that Millie built. At my slightest encouragement, memories poured out of the doctors and nurses who work there: Millie's intense, unmatchable dedication. The way her fellows were her family. The Christmas parties when she served spiced wine spiked with pure grain alcohol and held mistletoe shoots, which involved harvesting sprigs out of the mistletoe tree using a shotgun. And, most of all, the "Millie rounds," in which she would grill interns about their tiny patients and throw the chart at their heads if they got a detail wrong.

Stahlman lives about twenty minutes from downtown Nashville. Corey Reese, Stahlman's close friend and biographer, drove me to meet Stahlman for lunch. On the drive, we passed the mansions of the society that Stahlman had left behind—stone and brick giants, looming Southern Gothics with sweeping lawns and horses—and the country club where her family were members.

And then down a narrow road that cuts through a wood, canopied with a cathedral of trees: Stahlman lives in a spacious log cabin, set in a thicket of dense forest that she bought years ago. She was waiting for us, sitting in an armchair, wearing crisply pressed slacks and a sweater, reading that day's newspaper. Her eyes were still a piercing blue, her face long, angular, and severe, a face that has held its own. Her long white hair was plaited neatly and tied back.

Stahlman was happy to reminisce about her lamb lab and her blood gas research. But she wasn't interested in self-celebration. She had just done what needed doing. I asked her if she ever thinks about all the lives that have been lived because of her work, and she said no, not really. She thinks some other doctor would have done it if she hadn't.

But they didn't. And she did.

Martha Aileen Stahlman Humphreys—that five-pound baby who became the beautiful twenty-year-old girl—now goes by her married name, Martha Lott. She is a neonatal nurse who works at the Vanderbilt University Children's Hospital NICU, the very place where Stahlman, who is her godmother, treated her nearly sixty years ago. "I felt called," she told me, explaining her career choice. Sometimes, under the right circumstances, she said, she mentions her origins to parents who are looking for hope. Her life started with a long-shot stint in an iron lung, and today she can treat one-pound babies with high-frequency oscillatory ventilators. "It's amazing how far we've come, what we can do these days," Lott told me. "It really is."

8

Dr. Maria Delivoria-Papadopoulos and the Rugged Machine

NEGATIVE PRESSURE VENTILATORS, the iron-lung type of respirator that saved Martha Lott, fell out of favor in the 1970s, replaced by positive pressure ventilation, which requires a tube down the trachea, inflating the lungs with pressurized air.

This didn't happen because it suddenly became clear that positive pressure was better than negative pressure. In fact, positive pressure ventilation was and is associated with lasting lung damage and disease because it is much more invasive. It doesn't mimic the natural muscle action of breathing the way that an iron lung does. But the iron lung–style machines were not practical: Most importantly, they made it very difficult to access the baby, as the whole torso was sealed in the machine. And it was not always possible to use them on babies smaller than two pounds, as a seal that small is difficult to maintain. And so positive pressure ventilation, the kind of support that Mira was given in the delivery room, became today's standard.

But first, someone had to be willing to try it. Here is where the idea was planted.

Athens, Agia Sofia Children's Hospital, 1958: A young doctor named Maria Delivoria was caring for children with polio in iron lungs. When the power went out—which was often—she was forced to choose one lung to pump by hand while the other children in the ward struggled, and in some cases died. She owned exactly one skirt and one blouse and ate one meal a day. She injected herself with caffeine to stay awake.

Five years later, having arrived safely in North America, what was it to her, really, to be called crazy by a bunch of twenty-something medical residents? She was ready to try the "rugged machine"—her description of the early positive pressure ventilator—because she was tired, so tired, of watching children die.

Delivoria-Papadopoulos is now in her eighties and a professor of pediatrics at Drexel University in Philadelphia. Her office is in a medical school building, down a warren of cinder-block hallways lined with animal research labs. There she works at a long table stacked with papers and a laptop on one end. Wall shelves are piled with books and papers and a small, jaunty Greek flag. Delivoria-Papadopoulos has a wide, animated, lined face, black hair that is often piled in a pouf on her head, and expressive, gnarled hands.

She invited me to sit down. Before I got a word out, she wanted to know my daughter's name. "*Mira* means *fate* in Greek," she told me. She asked to see a photo and pronounced Mira cute. She wanted to know if I'm crazy about her and if my husband is as crazy about her as I am. (I am. He is.) "And so, is she smart?" She asked this last question with the directness of a grandmother, the kind who might stop you on the street and tell you to put a sweater on your kid. But also with the underlying anxiety of a doctor who knows that surviving sometimes comes at a price. "Yes," I said. "She's smart."

In 1960, Delivoria immigrated to Wichita, Kansas, with Chris

Papadopoulos, who would become her husband. She spoke no English, but soon she was ensconced in an internship at what was then called St. Francis Hospital, rotating through the OB ward, where she learned her first English word: "Push!" She was miserable in Wichita. She loved the nuns who ran the hospital, but it was a shock to go from war-torn Athens to a small city in central Kansas. She didn't understand her neighbors. "They kept saying 'How is the weather?' And 'This weekend, we'll do a picnic.'" She made a face like it was the most inane thing she'd ever heard, and I made a mental note to avoid small talk.

She got a residency at SUNY Downstate Medical Center in Brooklyn, where she felt more at home. She and Chris, who was also doing a residency, got one day off every two weeks and spent it at the Metropolitan Museum of Art. At SUNY she worked for two doctors, Alexander Wiener and Irving Wexler, who had discovered the Rh factor in newborns—basically a mismatch blood condition in which a mother's antibodies attack a fetus's blood, causing either stillbirth or the baby to be born (often preterm) with life-threatening jaundice and anemia. Now there are medications that a pregnant woman can take to prevent the reaction from happening, but at that time there was only one cure: to do an exchange transfusion after birth, essentially swapping all of the baby's blood with donor blood by removing a small volume of blood and then replacing it, repeating the process multiple times. (This is the procedure I had when I was born at 32 weeks.)

The exchange transfusion was done through the saphenous vein in the leg, which was not ideal because it could clot. Deliforia-Papadopoulos remembered one baby in particular. "Here is this cute baby," she said. "The poor baby was very sick. My role was to transfer the baby from the nursery to the doctors and to hold the pole while the blood was given." She started the transfusion, got halfway, and the vein clotted. "They had taken blood from the baby, but they could not give it back. The baby was this color." Delivoria-Papadopoulos rapped one gnarled index finger on a blank sheet of paper.

Wiener told her to take the baby back to the nursery and call him when he died. She wheeled the baby back into the nursery, where the nurses demanded that she try something—anything. "I looked and looked," said Delivoria-Papadopoulos, remembering her search for a suitable blood vessel. And I found, right here, this point." She tapped the center of her forehead. "I put the tiny needle in, and in it went. The baby turns pink, starts kicking around, crying. And the rest of us were celebrating." Later in the day, Wiener came by and asked why she didn't call him when the baby died. "I said, 'Well, the baby didn't die,' and I explained what had happened. He was pleased." The next day the baby's father came in and got down on his knees to her in gratitude.

That baby was as important to her as she was to him. She had always known she wanted to be a pediatrician, but now she saw a new, specialized field emerging, and she wanted in. Neonatology was full of possibilities. It was full of driven iconoclasts like her, people who were unwilling to accept the dismal status quo. "After I could do that for that baby, I said, 'I will spend the rest of my life with babies,'" she told me. "That's what made me go to neonatology."

After her residency in Brooklyn ended, Delivoria-Papadopoulos moved with her new husband to Toronto, where she got a fellowship in neonatology at the Hospital for Sick Children, under Dr. Paul Swyer. As was the norm at the time, there was not much there in the way of treatment for RDS, just supplemental oxygen piped into the incubator. Babies died routinely.

Delivoria-Papadopoulos had experience with artificial respiration for the treatment of polio, and her interest was piqued by the positive pressure ventilators that had recently become available for adults and older children during surgery.

Delivoria-Papadopoulos thought they could adapt the adult ventilator, an early model called the Bird Mark 8, for use in premature babies. It was a good place to start, at least. "I see the babies dying right and left," she said. "I asked for a ventilator to help me make the babies breathe." The residents, in turn, asked her if she was crazy.

The residents at that time were responsible for each patient's care. As a fellow, Delivoria-Papadopoulos was more experienced and technically higher up, but she was not part of the same patient care chain of command; she was a researcher, more like a consultant. It's true that Delivoria-Papadopoulos's proposal was untested in the extreme—only a handful of doctors worldwide had tried or even suggested it—but none of the residents were offering any bright ideas, and meanwhile the babies kept dying. It probably didn't help that Delivoria-Papadopoulos was a woman and spoke with a Greek accent—the residents weren't accustomed to taking direction from someone like her.

Luckily, she didn't particularly care what they thought. But first things first: She didn't have tubes small enough to connect the babies to the existing machine. Delivoria-Papadopoulos wrote to a friend of hers who was a respiratory physiologist and he came up with two tiny tubes, which he mailed to her.

"I had the tube. I had the respirator," she said. "The resident says, 'That's our patient; you cannot touch them.'" Delivoria-Papadopoulos shook her head. "They said, 'You are experimenting on our babies.' So I said, 'Okay, I will show you when a baby dies.'" This was the plan: wait until death, then try the ventilator.

So a baby was pronounced dead, and Delivoria-Papadopoulos had the ventilator ready. The residents made her wait for five minutes after death before they would let her near the child. "And now I intubated the baby," she said. "I put them on a machine and they revived for an hour. They revived after being pronounced dead."

She did this over and over and over, threading tubes into cold little bodies, restarting hearts that had gone still. Of those first babies, she had no survivors—the babies were too far gone; the machine was primitive compared to what we have today—until finally the residents started letting her intubate at the moment of cardiorespiratory arrest, when death would have been pronounced, rather than making her wait several minutes.

Talking about these babies almost sixty years after the fact, Delivoria-Papadopoulos still remembered them as individuals and wept as she spoke about them. What she has seen in her life, from Athens to Toronto to Philadelphia, is enough to make most people numb. But she is not numb; these deaths were unacceptable to her. She felt the babies' lives were asking questions that she couldn't ignore: *What is best for them? Who should decide?* This is what allowed her, maybe forced her, to do what she did.

"Every baby, I had the hope that he will be saved. And after each death, I was very upset," she said. Finally, after one particularly bad day, when a baby girl survived for four days on the vent and then contracted pneumonia and died, she decided she would quit. She told Swyer that she was done with ventilation.

"It might have been two days later," she said. "And I'm being paged. I ran up to the nursery and there was this baby in cardiorespiratory arrest. I forgot my promise. I intubated the baby." The girl, who had been born at 34 weeks, was on the ventilator for seven days and went home after a month and a half, the first survivor.

Just a few months later, in May 1963, Deliviora-Papadopoulos presented her research to the Society for Pediatric Research in New Jersey. She had data on eighteen babies (although she had intubated many more than that) and among them the one survivor. She was nervous. But her presentation was met with excitement. "A baby that was presumably dead survived," she explained. She had successfully treated respiratory distress syndrome in a premature baby with positive pressure ventilation, something that had been thought impossible. It was a turning point. In 1965 she and Swyer published more results, reporting seven survivors. It was an acceleration. Someone else had taken the risk of going first. Other doctors started to be willing to try this new, rugged machine on the babies in their care.

In a way, like Stahlman, Deliviora-Papadopoulos *was* experimenting on the babies: There were no randomized trials, no FDA approvals, no formalized parental consent. For the babies who were intubated after cardiac arrest and brought back, only to die from infection or a burst lung or a brain bleed, it was not a kind death. It is hard to think of them being denied everything, even rest.

Dr. John Lantos has written extensively about the history of neonatology. "It was incredibly morally complex," he said of Deliviora-Papadopoulos's early work. "But everybody in the world would have said there's nothing we can do to save these babies. Little Maria—what was she, twenty-four years old or something?—an immigrant from Greece, decides she's going to go in all by herself and try to save them. Pretty amazing. And changed the world. There's probably half a million people in the world today who wouldn't be here but for her work."

As the years went by, the ventilators improved and so did

Delivoria-Papadopoulos's outcomes. In time, positive pressure ventilation would become the standard practice for treating babies with severe respiratory distress.

Delivoria-Papadopoulos went on to establish the first NICU at the University of Pennsylvania, where she is a professor emeritus, and later directed the NICU at Drexel University, where she still teaches and does research on how to prevent and treat lack of oxygen to the brain at birth. Her husband died a few years ago. Her office is covered with photos of colleagues and of her two sons.

Before I left Delivoria-Papadopoulos's office, there was a contemplative silence between us. And then she said: "Now there's zillions of NICUs. Babies are ventilated. Babies leave. The machines became much, much, much better. Nothing like the rugged machine I had. Some of the babies who would die, it was because of the machine. It was not smooth. But today it's a poem. You know."

I do know.

Parents wait to hear that first breath, the cry that announces that their baby has successfully made its way into our world. Its absence, the terrible silence following Mira's birth, meant that it didn't feel much like a birth at all. It felt like she was suspended between life and whatever comes before it. As she lay limp and blue under the warmer, her future was dependent on the neonatologist's skill at placing the tiniest tube into her mouth, on his ability to thread this tube between her vocal cords, down her trachea, to sit just above her lungs, which must have been the size of nickels. And then he connected that tube to a ventilator, which gave her what she needed: breath.

Today this procedure is relatively routine in neonatology, a skill that must be mastered, passed down from doctor to doctor. But there is an invisibly high price behind it because it was gleaned from other babies: the technique, the tubing, the machine—all products of decades of trial and error.

Later, in the middle of the night on her first day, an X-ray revealed Mira's right lung was partially collapsed, with areas throughout her lungs that appeared granular and opaque, a hallmark of RDS. None of the clinicians were surprised or even particularly worried. Fifty years earlier it would have been a death sentence.

So if endotracheal tubes and pressure calculations and oxygen concentrations can be a poem, then this is a poem. At the time, I felt all the tubes and wires penetrating her tiny body were a violation, though a necessary one. Now, when I look at those images of Mira, red and emaciated, just born, the endotracheal tube snaking from her mouth, I still flinch. But I also see Stahlman and Delivoria-Papadopoulos, and I see Alexander Graham Bell's two boys, and all the anonymous babies who didn't make it, who never even had a chance. And I take a breath.

9

JFK's Lost Baby and the
Advent of Surfactant

ON AUGUST 7, 1963, Jackie Kennedy was 34 weeks pregnant; that morning she took her daughter, Caroline, to ride her favorite pony, Macaroni, in Osterville, near the family's compound in Hyannis Port, Massachusetts. A warm, clear Cape day. When the pains came, Jackie knew what they were. Two hours later, Patrick Bouvier Kennedy was born by emergency C-section at Otis Air Force Base in Bourne. He weighed 4 pounds, 10 ounces: an instant celebrity, the most famous premature baby who ever lived.

He was a sweet-looking boy: cute, fine featured, with a glossy thatch of light brown hair on his head. But almost the moment the cord was cut, his tiny chest started heaving painfully, rapidly. Instead of a recognizable newborn cry he emitted a high-pitched whimper and grunt. He was baptized immediately, then placed in an incubator, and oxygen was piped in, in increasing concentrations, despite the risk of blindness. Nevertheless, Patrick breathed faster and faster and his skin took on a blue cast. Jackie's obstetrician, Dr. John W. Walsh, knew that he had reached the limit of what he could provide at the small military hospital on Cape Cod. He called Boston Children's Hospital, and soon Dr. James E. Drorbaugh, a pediatrician, navy veteran, and father, was on a helicopter, flying south to Bourne.

President John F. Kennedy had been notified when Jackie was in labor; he left the White House immediately, where he had been in meetings negotiating the Limited Nuclear Test Ban Treaty. From there, he went to Andrews Air Force Base and flew to Otis, where he arrived less than an hour after Patrick's birth. Drorbaugh and Walsh met the president and told him that the baby needed to be transferred to Boston Children's Hospital, where more, perhaps, could be done.

First, Patrick was wheeled into Jackie's room, where she was coming out of sedation, and one porthole was opened so she could stroke his head for a minute—but the oxygen in the incubator had to be conserved. Jackie was distraught, and had been since she realized she was going into early labor yet again. She had already had one stillbirth, a girl named Arabella, and one miscarriage. Caroline had been born full term, but John Jr. had been four weeks early and had needed incubation and oxygen for milder breathing difficulty.

Before the president left, he directed that the television in Jackie's room be disabled because he didn't want her to learn of the baby's condition on the news. Everyone downplayed the situation to Jackie, trying to keep her calm. John had been in the same boat, they said, and look at him now.

By this point in 1963, both Delivoria-Papadopoulos and Stahlman had documented survivors. Patrick would have had reasonable odds to join them if he had had access to mechanical ventilation. His condition, in size, gestation, and severity of his RDS, was quite similar to that of Martha Lott. Unfortunately, the institution where he was going, Boston Children's Hospital, did not offer mechanical ventilation for newborns.

You would think the Kennedys would have the best of everything, and indeed, Boston Children's, affiliated with Harvard, was on the forefront of pediatric care, as it is now. But mechanical ventilation at that time was still considered entirely unproven, more likely to cause a burst lung than to be an effective treatment. And it wasn't as though a doctor could just order a respirator for a baby from the medical equipment company: Remember that Delivoria-Papadopoulos had to tinker with an adult machine to adapt it for babies, and Stahlman got one of only a handful of mini–iron lung prototypes. Boston Children's had an evidence-based, gold-plated reputation to protect. Mechanical ventilation for babies was just too out there for them at that time.

As the ambulance sped northward from Bourne to Boston, a police escort clearing the way, crowds of onlookers gathered on the highway overpasses; word had spread through the White House press corps that Jackie had given birth, the first child born to a sitting president in sixty-five years. And a Kennedy! Patrick arrived in Boston around 7:00 p.m. that day, and by the time his father arrived a few hours later, there were throngs of well-wishers outside.

Drorbaugh was tasked with Patrick's care. It was quite obvious to him and everyone else who set eyes on the baby that Patrick was not doing well. He skin remained gray-blue; he panted, his chest retracting with each breath. Like Stahlman, staff at Boston Children's had the rare ability to measure blood gases from an umbilical catheter, and what they saw in Patrick's appearance was borne out by his lab results: very low oxygen, high carbon dioxide, and blood that was too acidic, which is a side effect of lack of oxygen. So they piped high concentrations of oxygen into his incubator—the long-term risks were not worth thinking of—and gave him intravenous fluids and bicarbonate, which helped lower the pH of his blood and keep it from getting increasingly acidic. Patrick seemed to stabilize, or at least he didn't get worse.

The public was transfixed. The next morning, August 8, newspapers around the world led with the story of the newest Kennedy's birth and his uncertain condition. The morning edition of the *Boston Globe* covered almost nothing else, with a towering headline: "Baby Sped to Boston: Has Trouble Breathing; Kennedy Stands by Here." In that story, Pierre Salinger, the White House press secretary, denied that Patrick was on a "danger list," which was an odd but clear way to phrase it. Just below the fold, another headline that reads, "He's a Kennedy—He'll Make It," now has a sadly inaccurate ring. And *Globe* reporter Gloria Negri wrote an ode to the baby who had brought the whole world to a halt:

"He is only 4 pounds 10.5 ounces but Wednesday he held the heart-strings of the world in his little red fists. In Washington, Bonn, Squaw Island, Hyannis Port, Rome, Paris and in the Kremlin, heads of state waited upon his every move along with the peasant in the field and the man on the street."

That afternoon, Patrick's condition worsened, and the president returned to the hospital. The baby was awash in 100 percent oxygen in his incubator, and it wasn't helping. The doctors had run out of options.

Despite the extensive press coverage, there are few public details from the clinicians who actually treated Patrick, who, despite being a Kennedy, was entitled to privacy. *Patrick Bouvier Kennedy: A Brief Life that Changed the History of Newborn Care* by Michael S. Ryan contains the most detailed (and perhaps the only) interviews with Drorbaugh and two other consulting physicians. When I contacted Drorbaugh, who is now retired and living in Hawaii, he declined to be interviewed, but his daughter referred me to Ryan's book.

In the 2012 interview with Ryan, Drorbaugh mentioned that Boston Children's was swamped with outside offers of clinical advice and expertise. He implied that he didn't accept any of the offers and that he himself never reached out for assistance.

Janet Auchincloss, Jackie's mother, took her own initiative. She quickly tracked down Dr. Samuel Levine of Cornell in New York, because he had successfully treated her granddaughter Anna Christina, who was born to Lee Radziwill weighing only three pounds. She had Levine snatched from his apartment on the Upper East Side and flown to Boston, where she insisted that he be consulted. But there was nothing additional he could offer.

Both Delivoria-Papadopoulos and Stahlman say that someone called them and asked them to bring their ventilators and come help the president's baby. Both had essentially the same answer: *I can't leave my patients, because without me, they'll die.* At that time, there were not cadres of physicians and nurses who knew how to use the equipment. The advent of true NICUs involved training and systems that ensured that the care wasn't predicated on the skill and availability of just one or two people. But at the time it was really just them at the bedsides, day and night. And their machines were not very portable. Both said they would have gladly treated Patrick if he could be transferred to their hospitals, but his condition was too tenuous for him to be moved.

Delivoria-Papadopoulos said that she remembers it like it was yesterday. Stahlman still sounded emotional talking about it, sitting in her cabin fifty-four years later. She wanted to help, but what could she do? "We were busy, busy, busy," she said, shaking her head, remembering the impossibility of dropping everything to go to Boston. "Whose baby is more important than anyone else's? Is the Kennedy baby more important than your baby? Not in my book."

It's hard to untangle the exact chain of communication here: It was an emergency; the family was desperate; everyone wanted to help; a child's life was on the line. I can understand why no one at Boston Children's would have wanted to admit, even years later, that they were asking for help. Or maybe it wasn't Boston Children's calling at all; it could have been Kennedy family and friends reaching out for second opinions, other options, working behind the scenes.

Drorbaugh said later he considered mechanical ventilation unacceptably risky for Patrick. But considering what he agreed to next, with his back against the wall, it seems plausible that he might have tried ventilation if he had had the equipment available.

In consultation with multiple higher-ups, including Dr. William F. Bernhard, a pediatric heart surgeon, it was decided to put Patrick in a hyperbaric chamber, the same kind used to treat the bends, which Bernhard used as an operating room for babies with heart defects. He had successfully performed about one hundred cardiac surgeries inside the chamber, which had just enough room for several clinicians and a small patient. The idea was that the high pressure made oxygen dissolve more easily in the blood, thereby temporarily increasing oxygen saturation. It was not at all intended to treat RDS in premature babies. Bernhard, in his later interview with Ryan, said that putting Patrick in the chamber was a desperate move, motivated by the feeling that they were losing the president's baby. It was not a proposal he would normally have agreed to. Then again, who could say? Well-informed desperation has been known to produce new treatments.

Heroic measures must have seemed necessary. The whole world wanted a happy ending. The press was gathered outside. President Kennedy was pacing the lounge. Jackie, still in her hospital bed on

Cape Cod, had lost two other babies. Doctors advised the president that Patrick was not doing well but that they had one last-ditch idea.

They set the exhausted child into the tank, an iron behemoth built by the navy in 1928, which was in the basement of the hospital. Drorbaugh, along with an anesthesiologist, a technician, and Bernhard, filed into the cramped white-tiled space. They set it to three atmospheres of pressure, the equivalent of being about sixty-five feet underwater. The president, joined at various times by his two brothers and Levine, watched from behind thick portholes.

For a little bit, as the chamber pressurized, Patrick's condition seemed to improve. And then it didn't. Almost twelve hours later, still in the pressurized tank, the baby was using every muscle in his body to gasp for breath. The end stage of respiratory distress, as Ryan describes it in his book, is difficult to read and must be harder still to watch, an agony if it is your child. Often the child lies flat and still, covered in sweat. Their arms and legs are floppy; their skin is gray, and the only movement is that of desperate attempts to breathe: The nostrils flare, the throat hollows with each intake of air, the chest heaves at an impossibly rapid rate.

The doctors inside the chamber were able to speak to the president and others outside through a two-way intercom. They told him they were losing the baby. After Patrick stopped breathing entirely for a bit, the anesthesiologist started bagging him, trying to squeeze air into his collapsed lungs. It was not going to work. At 4:00 a.m. on the morning of Friday the tenth, Patrick gasped for the last time. He had been fighting suffocation for thirty-nine hours. President Kennedy walked away to cry; Auchincloss, Jackie's mother, later said it was the only time she saw the president so undone. The doctors inside the chamber with the baby—the baby who was finally still—had to wait several hours for depressurization before they could open the door. By that time the president was gone.

The next morning, the *New York Times* ran an op-ed titled "A Little Boy." It read in part, "The world watched the pathetic and yet stirring fight for the life of little Patrick Bouvier Kennedy with a special sense of human sympathy and of understanding for the travail of his parents. And now that he is gone, we mourn with them the loss of their little boy."

I can't help but imagine an alternate, impossible, timeline: What if Patrick had stayed on Otis Air Force Base with Jackie and died peacefully? He was already getting incubation and supplemental oxygen, and Boston didn't have much more to offer than that. It's true that the pediatricians at Boston Children's were able to obtain his blood gases—an amazing innovation at the time—but those lab tests only served to confirm what everyone could see with their eyes.

By all accounts, the family wanted every possible treatment. They had already had a similar experience with John Jr., so they had reason to believe treatment would work. In the end, they were just parents hoping against hope. And the doctors' intense desire to help the president's baby—anyone's baby—is deeply understandable. But what if they had been able to see things differently? Bernhard said in his interview with Ryan that he never put patients with lung problems in the hyperbaric chamber because he knew it wouldn't benefit them. Nevertheless, he was swept along with the momentum: the best pediatricians in the country, the best hospital, the best university. Surely there was something they could do? In reality, the only people who could have done something were John and Jackie Kennedy: They could have held their exhausted little boy.

The sad, short drama of Patrick Kennedy's life was eclipsed by the murder of his father just three months later. But Patrick's very public struggle set powerful wheels turning. Suddenly there was wide awareness of this problem, and not only that: The most admired family in the country had been directly affected. Drorbaugh recalled getting a letter from President Kennedy's press secretary shortly after Patrick's death, asking for guidance about how to set up special funding for RDS research. This led directly to an NIH push to find treatments for premature babies' immature lungs.

Where there is funding for research, there is research. Suddenly it wasn't just Delivoria-Papadopoulos and Stahlman and a handful of others laboring in relative obscurity: Many institutions wanted to solve the problem that had stolen the last Kennedy baby.

First came the improvement of artificial ventilation. By 1970 the inventor of the Bird ventilator had come up with a model specifically made for newborns, called the Baby Bird. It was mass-produced and became the first neonatal ventilator at many hospitals.

At the same time Patrick's death led to a race to synthesize pulmonary surfactant, that mysterious substance whose absence is the main cause of RDS. Surfactant is a slippery foam, a mix of lipids and proteins, and it turns out to be extremely difficult to make in a lab. Dr. Marshall H. Klaus, an early neonatologist working in California, tried to figure out its composition by going to a slaughterhouse each week and getting a couple of cow lungs. He'd spin the material he found in the lungs to extract the lipids and, by examining that, devised a synthetic surfactant. In 1967, he and other colleagues, including Dr. Jacqueline Chu, tested this aerosolized concoction on babies with RDS in Singapore. It didn't work.

The quest would take more than another decade, until the early 1980s, when researchers in Japan, Belfast, and Stockholm picked up the baton. Slowly it became clear what had gone wrong before:

The original experimental artificial surfactant had not included an important protein component. When purified natural surfactant was used and deposited directly into the lungs, it worked. "The white creamy mixture literally turned these blue babies pink within minutes," wrote one researcher in amazement.

In fact, the surfactants that are most effective even now are pulled directly from the lungs of pigs and cows and purified, an expensive, difficult process. For every step that we take into a future in which everything can be quantified and replicated, we remain tethered to these messy, intricate bodies. The biomedical companies that make surfactant actually employ people to go to slaughterhouses, insert tubes into just-killed cow's tracheas, and painstakingly extract the surfactant from one animal at a time. One set of lungs equals enough surfactant for two babies.

Many of the smallest babies are now treated with both a squirt of surfactant and ventilation of one kind or another. And it's rare for preemies to die of respiratory distress. Today, Patrick Kennedy would probably spend a couple weeks in the NICU and go home unscathed, to a wide-open future.

Part IV

The Self

Protecting the Premature Brain

10

The Revolutionary Practice
of Listening to Preemies

IF DR. HEIDELISE ALS were in charge of NICU design, the units would look almost nothing like they do now. Instead they would consist of individual "womb rooms," each kept at the temperature and humidity level of an incubator, an atmosphere that would be preserved by some sort of double-door air lock system so that people could go in and out without letting in cool, dry air. Inside, a baby would live with loved ones, ideally primarily the baby's parents, but also a rotating cast of grandparents, aunts, uncles, and close friends. Siblings would be welcome. Much of the time, the baby—who could still be attached to various respiratory supports, IVs, and monitors—would be snuggling skin-to-skin on one of these loved ones' chests. The room would be equipped with a double bed and recliners, a private bathroom. Nutritious meals would be provided. Each room would basically be a huge incubator, big enough for the whole family.

Als's vision, which she knows is unlikely to be realized anytime soon, reflects her belief that one of the most significant losses preterm babies experience is the deprivation of constant, gentle human connection. And she has found that that loss, and everything that goes with it, has consequences for brain development.

After Als told me about her theory of womb rooms, she turned her bright-blue-eyed gaze to me and asked me quite seriously what I thought of her idea, if I would have wanted this for Mira. I told her I would have loved it.

Als is a developmental psychologist whose research and advocacy have revolutionized the ways that babies are cared for in the NICU. She was born in Germany in 1940. Today she is a slight, silver-haired Harvard psychology professor who directs the Neurobehavioral Infant and Child Studies program at Boston Children's Hospital. She is perhaps the world's foremost expert on the developmental psychology and selfhood of premature babies and their bonds with their parents. She is a pioneer of what is now called newborn developmental care.

One outgrowth of Als's work is that private rooms (though not womb rooms) are increasingly the norm in newer NICUs in the United States and in other high-resource countries. In Shreveport, Louisiana, one summer afternoon, Dr. Gerald Brent Whitton was showing me around the Willis-Knighton South's Center for Women's Health NICU, which was renovated and expanded in 2007. "What do you notice?" Whitton asked me, pausing for a moment in a private room with an unoccupied incubator. The room was spacious, more than big enough to comfortably fit three adults, along with the incubator and an upholstered reclining chair for kangaroo care. A large window that could be covered by a heavy shade took up one wall, and on the opposite side was a small nurse's workspace, with a monitor, a computer, and a desk chair. This is what most new NICUs look like, and they are quite different from the older, incubator-on-incubator-on-incubator open ward where Mira spent her first months.

But the most striking difference—the one Whitton wanted me to notice—couldn't be seen, only heard. Or, rather, not heard. "It's so quiet," I said. "It doesn't sound like a casino, right?" he asked. It didn't. The air had a muffled quality. With the door closed, the alarms from the other rooms were inaudible, although each nurse could still keep tabs on their other patients through monitors and wireless devices that looked like cell phones. This was infinitely more pleasant than being in a small room filled with six or twelve constantly beeping incubators, the periodic rush of an emergency, the intimacy with other new parents' most fearful moments: An alarm shrills and six mothers' heads snap up. *Is that mine?*

But private soundproof rooms aren't the trend because they are more relaxing—or not precisely. Such rooms reflect relatively new understanding about how harmful sensory overstimulation can be for preemies' delicate and developing brains. And parents tend to spend more time in the NICU when they have a private room. Generally, the more family involvement, the better the baby's outcome. That means that these rooms can actually result in healthier babies in both the short and long term.

When Als started her groundbreaking work in the 1970s and early '80s, this level of attention to a baby's experience was unimaginable. In fact, preemies in the NICU were thought to be primitive creatures, impervious to the world, because their brains were so immature. They generally weren't given pain medication or anesthesia for painful procedures, even surgeries. (In fact, it was not until 1987 that the American Academy of Pediatrics declared that babies in fact do feel pain and should get anesthesia and pain relief to the same standards as other patients.) Clinicians handled the infants often but parents were either not encouraged or not allowed to visit, to hold them, or to touch them. The babies lay in noisy incubators flat on their backs on hard surfaces, covered in plastic wrap to keep their skin moist, splayed out like little raw chickens under bright lights. Sometimes they were sedated or restrained, their arms tied down so that they couldn't pull out their lines and tubes.

Nevertheless, in the context of the time, this care was increasingly effective; respiratory support was getting better and better. Babies who would have died in earlier years started to live in increasing numbers. But after spending their first months uncomforted, pinned down in the glare and hustle of the NICU, many of these babies' neurological outcomes were not good.

Felice Sklamberg is an occupational therapist who works at the NICU follow-up program at New York University; I met her when she took care of Mira, and she later told me that Als was her professional inspiration. She remembers the days before Als's research, before developmental care. "When those kids came back to follow up, they had flat heads, shoulders hunched up, legs like Charlie Chaplin, stretched by gravity," Sklamberg remembered. "A lot of cerebral palsy, visual impairment from too much oxygen and the lights. Much more impaired than they are now. Neurologically devastated. Heavy-duty motor, cognitive, and visual handicaps."

In these earlier NICUs, survival was the goal. No one meant to be cruel or cause harm. It was assumed that premature babies were not sentient or capable of much human experience: pain, comfort, avoidance, communication. Maybe the clinicians had to think that, in order to do what they had to do. They were only just getting good at treating these children, and there was little understanding of how survivors' brains might be affected by their NICU experience—that even the survivors who did well might have permanent differences in their brain function. But of course it makes sense: We did not evolve to gestate this way.

The brain development that makes us uniquely human is accomplished in the last part of pregnancy. Or, for premature babies, it is accomplished in the NICU. Between 28 weeks and term, the fetal or premature brain triples in weight.

The human brain forms from the center outward. The brain stem, which controls reflexive and involuntary activity like breathing, heartbeat, blinking, swallowing, and sucking, is one of the first structures to mature; nevertheless, you only have to spend a few hours in a NICU to see that many of the babies' brain stems are not quite up and running. Until 34 or 35 weeks they may have difficulty coordinating sucking, swallowing, and breathing, which is why they are often tube fed. They may often have episodes in which they don't breathe, or their hearts just slow toward a stop. As they near term and their brains mature, those episodes usually peter out.

The cerebral cortex, the outer layer of the brain, is the last brain structure to develop and mature; in premature babies it is still very much under construction. The cortex handles our abilities to take in the sights, sounds, and smells of the world, to communicate with each other, learn, plan, and remember, to have an inner life. The cortex's awakening brings about what neurologists call "eloquent brain function," a lovely way to put it.

When you think "brain," what springs to mind is probably the cortex: those wrinkly gray furrows. It is only about 2 millimeters thick, but its vast potential comes from a large surface area, intricate gyrations that evolved to pack more surface area into a small space and therefore more neurons, more connections, more complexity. But if you peered inside a 24-week baby's head, you would see an organ that has very few furrows, very little gray matter; like a smooth white bean. The gray convolutions have yet to grow.

And exactly *how* we humans accomplish this task is remarkable: The neurons form in the center of the brain and then they move outward in conveyer-belt fashion, migrating into place, stacking themselves up into the most complex biological system we have. This migration starts around week 12 and mostly ends by week 28. If the neurons don't stack correctly, it can lead to various issues, from cognitive problems to epilepsy. Once in place, these neurons start to organize themselves, sprouting branches to communicate with each other through synapses, the connections that allow our brains to work. Myelination, the formation of protective fatty sheaths around the neurons' branches, accelerates around 34 weeks. Myelin enables electrical impulses to be transmitted from neuron to neuron quickly and efficiently. Preterm birth can disrupt myelination and leave children vulnerable to motor and cognitive problems.

In a very literal sense, premature babies are engaged in a delicate process of building themselves. They are caught out in the open, unprotected, in the middle of this complicated project, which they must continue under difficult conditions.

Nevertheless, they *are* capable of communication, of feeling pain and stress, of being comforted. But preterm babies' selfhood and *way* of communicating, their relationship to their parents and to their environment, is different from that of term babies, because they can't deal with sensory inputs the same way term babies do. You might intuitively know how to cuddle, rock, and speak to your term baby. But you have to *learn* how to interact with your preemie: calmly, quietly cradling them against your skin in a way that makes them feel gently contained and comforted, not rocking or stroking or bouncing.

"Intuitive parenting for the full-term baby is regulated by evolution," explained Als. "There are exceptions, but it works in the majority of cases. For these preemies, we need to be more aware and attuned so that we understand the baby and don't overstep him."

Just for instance: Term babies tend to like black-and-white patterns. They can look at the pattern and then, when they have had enough, are able to look away. They are able to regulate their own attention and keep themselves in balance. A baby who is, say, 32 weeks' gestational age can focus on a close-up black-and-white pattern but lacks the ability to look away when they have had enough stimulation. They don't yet have the ability to both engage and disengage. So they'll fixate on the pattern and get more and more overstimulated, unable to reset their attention. Eventually their overwhelmed bodies may find a way to disengage: They might spit up or even stop breathing, or their heart rate might plummet.

Rushing preemies to do things that they don't have brain capacity for yet does not help and in fact can harm their later ability to process sensory stimulation. Given time, eventually that same premature baby will be able to look at the pattern and then look away.

What these infants need is revealed by what they would have been getting had something not gone wrong. Their systems are normal for where they are meant to be; their bodies have certain evolutionary expectations. Those expectations are for a life in the womb during the second and third trimesters:

You hear your mother's heartbeat, her muffled voice, gurgles from her digestive system, and faint sounds from the world outside. You can detect a bit of light and then darkness. You taste amniotic fluid; you feel your mother's hormonal rhythms and her sleep-wake cycles. You float weightless in warm fluid; you are gently contained by uterine walls that cradle you in the fetal position; it is easy to get your hands to your mouth; you can brace your feet against the muscular uterine wall. You sleep; you practice breathing. Everything is gentle, muted preparation. That's where the brain of a premature infant is meant to be.

Mature, neurotypical brains do so much for us so effortlessly: You can walk and talk to a friend at the same time, hardly even noticing the bright sunlight, the sound of the cars going by, the smell of the exhaust, and the feeling of the clothes against your skin. While you are busy living your life, the typical brain is humming along behind the scenes, filtering sensory input, preventing you from being bombarded by the world, allowing you to focus on what you choose.

It is an unfortunate double whammy for premature babies that intensive care units can be overwhelming even for a mature brain. Imagine that you wake up in a hospital bed with several loud alarms going off, fluorescent lights overhead, harsh cleaning fluid smells. You have several IVs in your arms, sticky electrodes on your chest, and a nurse is poking your heel with a needle to get some blood. You have a feeding tube down your throat and an oxygen mask strapped to your face.

Imagine experiencing all that with a brain that's not built to deal with anything but muffled sounds and faint light. Imagine having to develop the ability to filter sensory input while being assailed with sensory input.

There's an old saw in neurology that goes like this: What fires together wires together. Think of Pavlov and his dog. In some circumstances the brain can link, say, touch and pain. Eating and being startled. These associations can be built into babies' developing brains in ways that might affect the way they function later, the way they eat, sleep, move, talk, and feel.

Scientists think that this is a major reason why premature babies are at later risk of developing attention issues, learning disabilities, motor delays, and sensory processing difficulties. There is evidence that autism is more prevalent in former preemies than in the general population, but it's not completely clear why this is.

Some researchers think the disorder might have a distinct pathway, a different etiology, than autism in the general population—that this constellation of sensory issues might be directly linked to the experience of prematurity.

The difficulty is that what saves their lives is not good for their fragile brains. So the question is: How can we make life in the NICU as much like the womb as possible while also providing life support?

You can imagine that when Patrick Kennedy died, no one was asking this question. As long as so many premature babies were dying of respiratory distress, other concerns took a back seat. But as the 1960s came to a close, more and more hospitals were able to offer ventilation. In 1969, Als was doing her dissertation at the University of Pennsylvania on the communication loop between newborn full-term babies and their mothers.

Als was struck that everyone was always concerned with the mothers' behavior toward the babies, but she noticed that the babies were active participants in the relationship, too. "Who directs the interaction? Who starts the next sentence, if you will?" Als asked, speaking of nonverbal communication between mother and baby. "All the pressure was on the mother, the mother, the mother." But Als noticed that the babies could alter an encounter by sneezing or yawning, directing their gaze to their mother or away, and that there was a logic behind these behaviors. The babies were regulating their mothers—calming them, for example—as much as the mothers were regulating the babies. For instance, if a mother is holding her baby face-to-face, making direct eye contact and speaking to her, and if the baby starts to feel tired or overstimulated, she might sneeze. This interrupts the encounter. The mother will often then say something like "Oh, bless you" and tuck the baby up to her chest. "It really struck me that it is a two-way street," Als said. "The baby has a lot of power." In term babies, much of the time this communication loop happens instinctively and intuitively, one baby and one parent at a time, billions of times over.

It just so happened that Als was doing this research at Penn only shortly after Delivoria-Papadopoulos moved there from Toronto to establish Penn's first NICU. One day Delivoria-Papadopoulos noticed Als sitting in the nursery, observing the mothers and babies for hour after hour, collecting her data. In Als, Delivoria-Papadopoulos might have recognized something of herself: an uncommonly determined and empathetic scientific mind, a woman who grew up amid the chaos of World War II Europe. And so Delivoria-Papadopoulos said something to the effect of: *If you want to see interesting babies, come with me.*

Als remembers this first visit to the special care nursery, as the NICU was then called, as if it were yesterday. She watched the babies rushed in from the delivery room; first they would be bagged for respiratory support, and then some would be intubated and put on a ventilator. Doctors and nurses crowded above them; bright light shone directly on their faces. There were many hands, many tubes and needles. The babies were profoundly distressed by all this, but no one had the time to notice.

"It was very apparent that the babies did not want to be on the bags," Als said. "Before they were medicated, they would always fight to be on their sides, get their hands to the mouth, tuck their legs up. Of course, the doctors were much stronger and the babies would eventually end up being restrained, pinned down with their wrists and their ankles to not fight the breathing tube, what they all needed to do to help them survive.

"And it made such an impact on me, that as early as they came— probably around 28 weeks were the youngest—they were not some limp blob; they were babies. They were feisty, goal-directed babies. And yet after a few hours they lay there limply because they couldn't win that fight. They couldn't go back into the womb."

Als was somehow able to look through the emergency rush, beyond the question of survival. She saw the babies as individuals, and she saw that there was purpose and logic in their attempts to comfort and regulate themselves. She felt what they felt.

She was impressed with the nascent technology and the dedication of the clinicians, particularly Delivoria-Papadopoulos. But she wondered: Couldn't all this be accomplished without defeating the babies, thwarting them so thoroughly? Could they work with what the babies wanted and needed? From that very first NICU visit, Als started to develop the central thesis of her life's work, which would transform NICU culture: If we listen to premature babies, even though we don't intuitively speak their language, it will improve not only the babies' immediate medical condition but their long-term neurological and psychological functioning. This practice is now called newborn individualized developmental care. "They tell you what they need," said Als. "But you have to be willing to see it."

When the babies jerk, reach, and flail, they are searching for uterine walls. They quickly calm down if you use one hand to cup their heads and the other to cup their knees up to their bodies—the fetal position. If you place blankets or bolsters around them to approximate the soft containment of the womb and help them bring their hands to their mouths, their motor development proceeds more smoothly. If they grimace and yawn, open their hands wide and stretch their fingers, they are likely overwhelmed and asking for a break from stimulation. If they need a blood draw or a line placement, if a parent can hold them and speak softly to them, and their pain is appropriately treated, their stress response decreases. If you keep the environment around them dim and quiet, they have better sleep and better cognitive development, and their later brain functioning is less compromised. If you encourage parents to do kangaroo care, in which the baby rests on a parent's bare chest for body warmth and bonding, they do better by nearly every measure. If a baby goes very quiet and still, limp and pale, it doesn't necessarily mean they are calm; it might mean they are dangerously exhausted.

"What shocked me the most was the sensitivity of the preterm infant," Als said. She gave me an example: Say a nurse feeds a baby a dose of breast milk through a feeding tube. The baby's stomach distends and the baby, lying on her tummy, becomes overwhelmed. She goes limp. The nurse thinks: *Oh, good, the feeding made her calm and sleepy.* It's what you might think. But actually the baby is not calm. She simply has no energy to do anything except try to process that feeding.

After a few minutes the baby gains some strength back. She's attempting to do tiny little push-ups in her incubator to take the pressure off her abdomen and get on her side, but no one notices. In a uterus, floating, she could move easily, but here, under the weight of gravity, she is not yet strong enough. She gets exhausted. She throws up and stops breathing, and her heart rate plummets. The nurse comes back, says, "Oh, you spit up again," and turns up the baby's oxygen to help her recover. Now the baby has been overstimulated, is exhausted, has used precious calories, and has had her respiratory support increased, all because no one noticed her signals that she was uncomfortable. "We can watch the baby and we can prevent this," said Als.

Als's research and her methods are not just focused on the developmental health of the baby. She is also concerned with the family ecosystem, the parts inextricable from the whole. Since she started her career studying the dynamic between full-term babies and their parents, she is able to compare and contrast those experiences.

"When you are a healthy, full-term baby, when you open your eyes, your parent looks at you and talks to you, [and] you get more alert," Als said, describing typical, healthy interaction between term babies and parents. "That's not what preemie parents get. For a very sensitive, fragile little baby, everything has to be just so for them to open their eyes. Then the mother is so excited to see her baby's eyes. Then the baby gets overwhelmed and spits up."

Als explained how this miscommunication, this disconnect in the normal loop—what parents are evolutionarily programmed to expect from a newborn versus what they actually get from their preemie—can cause parents to have trouble feeling competent and bonded with their infant. It can even make a parent feel betrayed by their baby on some deep and unspoken level. And that, of course, can lead to guilt and shame. The way to avoid this, Als said, is to support parents: Explain that all this is typical of preemies, that there is nothing wrong with them or their babies. And to reassure parents that their baby needs them, empower them to read their baby's cues. With support, no one is more of an expert in understanding a particular baby than that baby's parents.

Als has decades of research detailing the neurodevelopmental realities of these experiences, but she actually learned how to be a baby watcher from her own son.

Als was born during World War II and grew up in a family of five; her father was an academic and a judge who avoided joining the Nazi Party but was nevertheless drafted into the army; because he had diabetes, he was assigned to work as a clerk in an army barracks. After the war, during the process of denazification, he did forced labor at a fish canning factory while he waited for his trial. He was ultimately found to have no culpability because he had tried to protect Jewish people from persecution when he worked as a prosecutor in the prewar years. He was reappointed as a judge in the new West Germany and would often let Als come to court with him, encouraging her innate fascination with human behavior.

Als remembered one dispute in their small Bavarian town: A man accused a farmer of allowing his cow to "surround him" and threaten him. Als told her father that she didn't see how a cow could surround or threaten anyone. Her father pointed out that it could *feel* that way—the huge animal, the whipping tail. He was known as a judge who sought compromise, who tried to help the feuding parties see each other's point of view. Als was fascinated by all of it, especially the process of trying to understand how other people perceived things. "I always watched people and made up little stories in my mind," she said. "Who I thought they were and what they were up against and what made them happy."

Als had graduated from college and was working as a school-teacher when she met a man from a well-off Philadelphia family who came to Germany with the U.S. military. He seemed to come from a charmed, cultured world, far from her family's worries about money and the future of their shamed and broken country. He pursued her, and finally she agreed to marry him.

They moved to Philadelphia, and in 1965 their son was born full term. But it was almost immediately obvious to Als that something was different about him. "Of course, I thought he was the most beautiful baby there was," she said. But she also noticed his left hand was always clenched, his left toes curled. She did not know this at the time, but these were signs that something might be neurologically amiss. He cried often and was very easily agitated. Als had to learn to interpret his subtle, nonverbal cues, like tuning in a faraway radio station. She noticed that when she kept the lights dim, when she moved slowly, he was calmer. She figured out tricks to keep him tranquil: distracting him with a mobile while changing his diaper, feeding him in a quiet place, just the two of them. As a toddler, he started having epileptic seizures, and he was eventually diagnosed with frontal polymicrogyria, an overfolding of the brain in the frontal temporal region.

Als doesn't know why or how her son acquired this brain condition in utero. But her experience with him, even when it was difficult, has fueled her desire to understand babies who communicate differently. She was—and is—struck by her son's tenacity and the resiliency of the mother-child communication loop: "He came to be who he is. He taught me to read him and adapt to him and help him get the next steps done," she said. The immensity of her love for him is matched by the intellectual curiosity that was born out of that love.

And so, as the sixties became the seventies, Als was raising a child with a significant disability and completing her dissertation for her PhD. In 1973, newly divorced, she moved to Boston to work with Dr. T. Berry Brazelton in his new Child Development Unit. Brazelton was a pediatrician and pioneer in infant development and communication. He was mainly interested in full-term babies, so Als set aside her fascination with preterm babies for a time, although they lingered in her mind. She worked with Brazelton on his Neonatal Behavioral Assessment Scale, an observational tool that helps clinicians quantify a full-term newborn's behavioral skills, neurological status, and potential vulnerabilities.

But she could not let the premature babies go. No one but Als seemed to see how much NICU care had to change. By day she taught research methodology to pediatric fellows. At night, when her son was asleep, she hired a babysitter to sit with him while she went back to the hospital, this time to the NICU, to watch the tiniest babies.

After years of sitting and watching premature babies and recording their behaviors in minute detail, Als started to codify and translate her observations into a clinical dictionary of sorts. Her guiding principle is to treat each baby as an individual. This is perhaps the hardest culture change of all, harder than making NICUs dark and quiet: Intensive care medicine likes protocols and rules and procedures that can be followed the same way for each patient.

Her approach, called the Newborn Individualized Developmental Care and Assessment Program (NIDCAP), is a system by which every clinician in the NICU can customize their care according to a particular infant's behavior in any given moment. Als and her NIDCAP colleagues are fond of saying that the baby doesn't—can't—fake it. They are always telling the truth about what they need. By observing the baby—Are they flailing and stressed? Sleeping and quiet? Looking around for a gentle interaction?—a caregiver can make an assessment using Als's system. What does this baby want right now? How can the necessary medical care be accomplished in the way that works best for the baby?

Dr. James Helm, an infant-family specialist who trained in the NIDCAP system under Als, directed the NIDCAP developmental care at the WakeMed hospital's NICU in Raleigh, North Carolina, for over thirty years. He said that the first thing that visitors to his NICU notice is how quiet it is—and not just in terms of decibel levels. He said it is a gentle place, a place where babies, not adults, set the pace.

"The activities are synchronized with the baby's response instead of 'Okay, now it's time for this, like it or lump it,'" he said, speaking of the way that mandatory care, like vital sign checks, are handled. "What we're trying to do is take that opportunity to support the baby with some wonderful human contact. Oh, and by the way, I just happened to slip in listening to your lungs, your heart rate, and your digestive sounds, and I've made sure that the IV wasn't infiltrated. We did all that stuff, but we basically had a nice time together, and now it's time to fill your belly and go back to sleep."

In Als's office in Boston she showed me a prototype of a positioning device that someone had made for her decades ago as she was developing NIDCAP. It was a pink knitted contraption that held a small doll about the size of a two-pound baby. The doll's legs were inside a soft knitted sack, and at the bottom of the sack was a firm, curved stuffed bolster that a baby could push against with their feet. Two strips of the knit gently crisscrossed the baby's torso, a kind of partial swaddle, and there was a hood at the top for the baby's head. Als undid the torso bands and held the doll to my chest to show how kangaroo care could be done without removing the baby from the sack. But then she showed me how the bands could be used incorrectly by being Velcroed too tightly. Her point was that these tools are important but not foolproof—not as important as the skill and care of the caregiver.

For instance, nurses sometimes flip a baby's tiny feet up toward their head to change a diaper. This can startle and upset the baby, and it can potentially change their blood pressure so rapidly that it puts them at risk of a brain bleed. Als suggested that there were slower, gentler ways of changing a diaper so that the baby hardly notices what is happening. Sometimes this involves "four-handed care," in which one caregiver, ideally a parent, comforts the baby with containing hands around their head and feet while someone else changes the diaper, takes the temperature, suctions a breathing tube, and so on. Obviously, four-handed care is doubly labor-intensive, especially if a parent is not available. But studies have shown that it results in a decreased stress response from the baby.

Dr. Elizabeth Rogers, an attending neonatologist and director of the NICU follow-up program at the University of California, San Francisco (UCSF), has recently opened a small baby unit, a cluster of six NICU rooms for babies who are born weighing fewer than 1,000 grams. The unit, called the Grove for its deep green walls, is focused on what is called neuroprotective care, a model that builds on Als's discoveries. Rogers says that the Grove came out of her desire to connect the dots between the challenges she was seeing in follow-up—like attention deficits—and the intensive care provided to those children at the very beginnings of their lives, which sometimes came with unnecessary stimulation.

Through her own research and that of others, Rogers noticed that increasingly advanced brain imaging has validated the core ideas of Als's work in a quantitative ("objective") form that has made it more acceptable to the medical mainstream.

"Als's work has been extraordinarily foundational to the idea of supporting a baby's development," said Rogers, "recognizing that every intervention we have impacts the developing brain, and learning to read the cues the baby is giving us in terms of how much they can tolerate developmentally, how much can they handle, while still managing their breathing and growing. Without Dr. Als's work, we would not have been able to do what we do."

In the Grove there is an emphasis on letting the baby rest when she wants to rest—clustering medical care together to minimize stimulation and sleep disruption—on watching for the baby's signals that she is ready for an interaction, on family-integrated care, on early kangaroo care when possible even for medically unstable babies, and on minimizing noise and light but encouraging parents to talk to and touch their babies in developmentally sensitive ways.

UCSF might be an outlier, though, in terms of taking these practices to heart and putting them into action; as an institution, it is more family-centered than most. The process of changing wider NICU culture to focus on the baby and family as much as on the professionals and technology has been a difficult, slow process, and it is still ongoing. Most NICUs have adapted to developmental care in some important ways: Encouraging kangaroo care for stable babies is the norm, as is covering the incubator, using soft positioning devices, and minimizing noise and light.

But Als's system is much more demanding and comprehensive. It is not just a matter of having good positioning pillows, or bringing in a physical therapist three times a week, or even just providing private rooms. Ideally, Als would change the way each nurse and doctor and respiratory therapist *thinks* before they ever open an incubator and lay hands on a baby: They would think of the baby as an individual, an integral and beloved member of a family, with preferences and needs—not just as a patient with a set of pathologies and a list of vital signs.

Felice Sklamberg, the occupational therapist who cared for Mira, started working at NYU's NICU in 1993, before they instituted meaningful developmental care. Sklamberg was following Als's research closely, and she knew change was necessary. Slowly she got the buy-in of the neonatologists and the nurses. She put heavy shades over the windows and muffling quilts over the incubators. She insisted that the decibel level in the unit be measured and not go over a certain threshold. She worked on positioning the babies in the incubators in more comfortable, natural ways in those little felt nests. She emphasized the importance of engaging parents in the care. These changes made a dramatic difference.

As the babies grew and came back to follow-up clinic as toddlers and preschoolers, Sklamberg saw the difference embodied. Before developmental care, many more kids had serious, life-altering disabilities. Now, she said, the majority of the kids she sees have more moderate or minor issues: developmental delays and learning disabilities that can be addressed with therapies, sensory quirks that can be accommodated. Kids like Mira.

One morning when Mira was about seven weeks old, I came into the NICU and found her with her eyes open, looking around. It wasn't the first time she had opened her eyes, but it was the first time I felt that she was choosing to look around, ready to engage. She was noticing what was around her instead of either sleeping or cringing from it. It felt like a powerful turning point. I thought: *There you are!* I understand now that there was a brain development reality behind that moment—that she had quietly accomplished the brain growth that she needed to be able to take more in, to be curious the way a baby is curious.

Als put it best: "It is totally amazing to me that we aren't in awe every time we step into a NICU, seeing all these little brains making the best of it."

11

Follow-up Care

Preemie Development Beyond the NICU

WHEN WE WERE discharged from the NICU, the doctors kept telling me that I should treat Mira like a "normal" baby. She didn't need intensive care anymore. I could cut it out with the surgical masks and the hand sanitizer. We had an appointment for the follow-up clinic in six months—a routine measure for babies born at her gestation and size—and she would continue to be seen there every six months until she was two years old. Other than that, we could assume this was all over. She was medically stable. We could go. But my anxiety level was still in the stratosphere. I knew the doctors were giving us accurate (if narrow) medical information, and that they were trying to be kind, but I didn't know how to act like she was a "normal" baby, whatever that means; I didn't know how to just let go of the two months of intensive care that were my introduction to being a mother. If Als had been there, she might have said something that would have made more sense: Prematurity doesn't end at discharge.

Development is an intricate cascade, a domino reaction, with each stage leading to the next, each skill building on the one before it. It seems miraculous to me now the way most babies figure out how to use their bodies in similar ways, a synchronized project that's almost unnoticeable because it looks so natural, makes so much sense, like an orchestra playing a symphony in which you hear only the song, not each individual instrument.

For premature babies, this cascade is disrupted at the very beginning. The last trimester of pregnancy, when a fetus curls more and more into the fetal position as uterine space gets tight, builds core muscle and coordination. It brings hands to the body's midline and to the mouth, setting up everything from nursing to grabbing toys. Being in an incubator, even carefully positioned with bolsters and with the best developmental care in the world, is not the same. It can lead to abnormal muscle tone (too floppy or too tight) and delayed motor skills, which can lead, in turn, to the delay of other skills.

Premature infants have to find their own cascade. As they grow, their milestones are on a different timeline. Being born early does not make babies jump ahead in development; in fact, it is often the opposite. Until they are two years old, "age appropriate" can mean that they are progressing as if they had been born on their due date. When a baby has the skills for what is called their adjusted or corrected age—that is, the age they *should* be—that's considered typical and not a problem. When they turn two, if they have not caught up to their actual age, they are then considered to be clinically delayed.

For example, when Mira was four months old in actuality, her corrected age was only one month—and it was expected that her skills would be that of a one-month-old. "Ooooh, how old is your baby?" from every stranger in the checkout line became an awkward attempt to explain that my eight-pound baby who appeared newborn was actually four months old.

Often, premature babies lag behind even their adjusted ages and need physical, occupational, and speech therapies to catch up.

Those developmental delays can simply be a matter of being immature and small, of spending your first few weeks or months in a hospital instead of in your parents' care, of using too many of your calories to breathe or gain weight rather than learn to hold your head up—the opposite of a head start. Delays and differences can also be the product of neurological issues, like those following from brain injury. If those persist, a child might be diagnosed with cerebral palsy, which simply describes a nonprogressive, incurable (though not untreatable!) developmental disability that arises from a brain injury at or shortly after birth.

I didn't know much about any of this when Mira first came home. I was still in the grip of an anxiety so vicious, I felt like a hand was curled around my throat. We were lucky, so lucky, so lucky. I felt as if I had only just escaped a tiger and my eyes were still squeezed shut. I obsessed over sudden infant death syndrome (SIDS), trying in vain to keep the temperature of each room at exactly 68 degrees Fahrenheit, the temperature a nurse mentioned as optimal for SIDS prevention. Mira's swaddle had to be a flawless, taut origami that came nowhere near her face; I would unfold and fold it over and over.

My main project in those days was trying to switch to breastfeeding directly, rather than pumping and bottle-feeding; breastfeeding, after all, supposedly helps prevent SIDS. Plus, I hated everything about pumping: Every three hours around the clock I was immobilized for forty-five minutes, tethered to this groaning, hissing machine that wrung the milk out of my blistered nipples. I hated sanitizing the pump equipment eight times a day. I hated finding room in the freezer for the bags of breast milk, and I hated remembering to defrost it. I hated my own guilt: I could not stop pumping; I had to give her this one thing.

So a lactation consultant came over to our apartment and watched me fumble with my big, floppy breasts and Mira's squirmy little body for a few minutes. She immediately broke the bad news: Mira's core and neck muscle tone was too low, and that was why we were struggling to nurse. She prescribed a battery of intricate neck and mouth exercises for Mira before each attempted nursing session, each of which was to be followed by a pumping session. (It was so impossible, I almost laughed in her face.) And she said I needed to get Mira into Early Intervention for physical therapy. Instead of being grateful for the heads-up, I was quietly furious. Mira had made it through the NICU. Now we were supposed to be done with systems and institutions and bad news. I was supposed to treat her like a normal baby! I heard "delay" and "disability" and I imagined disaster. I was ignorant of the tremendous spectrum of life that exists in those words.

With the help of Sklamberg, the occupational therapist who handles follow-up at NYU, we got an evaluation, and Mira did have low muscle tone (hypotonia), generalized weakness that led to motor delays, and torticollis, a preference for turning her head to one side, creating muscle tightness in her neck on the other side. This is also a common and treatable side effect of being in an incubator instead of floating in liquid. We started physical therapy, which mostly looked like gentle stretching in those early months. We never did figure out how to breastfeed.

When Mira turned one and was not crawling, her Early Intervention case manager told me that when babies can't crawl to explore their environment—grab and taste and smell stuff—their cognitive development can be affected as well. I hadn't appreciated how deeply motor skills influence other skills. *So what? She's slow to crawl! Who needs crawling?* Well, generally speaking, humans do.

There are, of course, other ways to develop cognitively: Not all babies crawl; not all children walk; it doesn't mean they can't explore and develop in other ways. But crawling isn't just crawling. It's an integral part of the cascade.

Shortly after that call, Mira did start crawling—clumsily, stubbornly, by leaning on her arms and shifting her right knee forward and then dragging the left leg behind her. Ideally, crawling would be symmetrical, so that one side of the body doesn't get strong while the other side gets weak. Her distinct and therapy-resistant preference for her right side over her left was *a thing*. Her wonderful Early Intervention physical therapist told us not to be too worried about it. Mira was just figuring out the ways that she could get her body to do the things she wanted it to do. But the asymmetry would occasionally lead one clinician or another to speculate that she could have mild cerebral palsy, which usually isn't fully ruled out or diagnosed until age two. But then, at seventeen months, she started walking, using both legs fairly equally, and those speculations petered out.

Mira's motor delays have persisted—she is four at the time of this writing—but they are all a matter of relative muscle weakness, not a neurological injury. She continues to be stronger on her right side than her left. At the moment, climbing and descending stairs is harder for her than for her peers, as is balance generally, which makes a crowded school staircase a bit of a safety issue. She has to work hard to use a marker or crayon the way her peers do, and she's only now getting better at feeding herself with a spoon or fork, as opposed to with her hands.

For her first three years, she struggled to gain weight, especially in the winter, when she was often sick. After landing in the ER for (terrifying) breathing difficulty a few times, Amol and I started bringing her to a pediatric pulmonologist, who diagnosed her with asthma, a common ex-preemie issue. Soon after, we realized that the inflammation in her airway and ear infections might have been contributing to her frequent vomiting. She got ear tubes surgically implanted to help her ears stay clear; with medication, her asthma became easier to control. As a result, she threw up less, and it was easier for her to gain weight, which in turn improved her motor skills.

Her challenges are minor in the scheme of things but significant enough that she still qualifies for both physical and occupational therapy and special education through the New York City Department of Education, and she is classified as a student with a disability. She may grow out of it—and even if she doesn't, it's unlikely to majorly impact her life. But the lesson I have taken from all this is that, developmentally, everything is connected.

To me, Mira is a wonder, the most determined person I have ever met. Luckily, she doesn't have to do it alone: The therapies she gets are imperative, because crawling isn't just crawling, walking isn't just walking, and holding a crayon isn't just holding a crayon. Physical and occupational therapies help her find the ways her body and mind can work best for her. And so they are imperative to her sense of herself as capable and strong, and getting stronger.

Every premature baby should get access to developmental care and therapies as they grow, if they need them. We were quite lucky in this regard: Not only did we qualify for our NICU's excellent follow-up clinic, but Mira also automatically qualified for Early Intervention because she had been born weighing less than 1,000 grams.

While NICU care is tremendously profitable for hospitals, follow-up care for survivors is not. Therefore, most hospitals want a NICU, but not all want to allocate money for a follow-up program. It is not only less profitable, it also garners less attention and prestige.

And some kids fall through the cracks: Not all NICU follow-up programs accept kids born after 32 weeks, although that later-born population accounts for 80 percent of all preterm births. In the world of prematurity, the later the birth, the better, in every sense. But that fact obscures something that is just as true: Compared to term babies, moderate and late preterm babies face much higher risks of developmental delays and disabilities. Yet they and their families may not get the ongoing support and monitoring that they need.

And, of course, not all families have insurance; not all programs accept Medicaid. In some parts of the country, families have to drive for hours to get any developmental follow-up care at all.

That is why a clinic like the Willis-Knighton Health System's outpatient pediatric rehabilitation in Shreveport, Louisiana, is so important. Within its region, it is one of the only NICU follow-up clinics that accepts Medicaid and it is the only one that does structured follow-up for all babies admitted to the NICU, not just those born before a certain gestation. This is especially crucial in Caddo Parish, where the clinic is located, which has the highest rate of preterm birth in the country, at 18 percent.

I flew into Shreveport, the largest city in Caddo Parish, on a hot August afternoon. The small plane bumped its way down through puffy clouds to reveal the densely wooded landscape below: a deep-green sweep of forest punctuated by swampy pools of still, blue water glinting in the sun. Lumber is a common industry in the parish, as is health care and natural-gas extraction, fracked from a Jurassic shale rock formation that lies beneath the region. It also has a thriving tourism industry, built around big jangly casinos on the banks of the Red River. And almost one in five babies is born too soon.

The number is so high—almost double the national average—that it is almost unbelievable. But it is the Louisiana Department of Health's own accounting, and it is consistent with March of Dimes' recent data. That rate of preterm birth is much more typical of economically exploited, nonindustrialized nations than of rich, industrialized countries like the United States. Caddo Parish's rate of preterm birth is higher than that of Haiti and Bangladesh; it is about the same as that of Malawi, which has the highest preterm birth rate in the world.

When I asked experts why this is, the answers usually centered on big, entrenched inequities like poverty, housing, nutrition, working conditions, exposure to pollution, racial discrimination, and classism. (These factors are what public health experts call social determinants of health.) Twenty-six percent of the population lives under the poverty line in Caddo Parish, more than double the national average.

The Willis-Knighton clinic is part of a strip mall complex of outpatient facilities—a breast center, a pediatric pulmonologist, and so on—just down the road from the Willis-Knighton South's Center for Women's Health, where one of the largest NICUs in Caddo Parish can care for up to fifty babies at a time. The NICU was entirely redesigned, renovated, and expanded in 2007, because the original twenty-some beds were not nearly enough for the number of babies in the area who needed intensive care.

About three months after each baby is discharged from the hospital, they come to the pediatric rehab clinic for an assessment of their development. And they continue to come, about every three months, until they are walking and talking well. Often that means until two years old, sometimes younger or older, depending on the child. (Routine NICU follow-up often lasts for about two years, but, unsurprisingly, longer is probably better, though not always possible with limited resources. Yale–New Haven Hospital now follows ex-preemies until they are four.)

Amy Sudduth is a physical therapist and director of physical medicine and rehabilitation at Willis-Knighton; she has been overseeing the follow-up clinic for over ten years. She said that part of the rationale for following all NICU graduates—not just those designated particularly high-risk by virtue of gestational age or birth weight—is that the clinicians are cognizant of the socioeconomic challenges in the community and the extra layers of risk that places on children. "A lot of these kiddos are born into not the greatest situations, so the more follow-up you can provide, the more opportunity you have to catch things," she said.

It was not yet 8:00 a.m., but the air was so humid that being outside felt like walking through fine, hot mist. Inside the clinic's pediatric gym, a group of five young women in scrubs and comfy shoes convened. They were occupational therapists, speech pathologists, and a physical therapist getting ready for what they call "NICU Fridays," because it's the only day they do these follow-up assessments. They will assess up to twenty-seven kids over the course of eight hours. Most of the kids were in the Willis-Knighton NICU, but some are referrals from other hospitals. Some live ten minutes away; others live hours away, in parts of the state that are extremely underserved. At least 40 percent of the children are covered by Medicaid.

Sudduth says that there is so much need in the community that she and her colleagues never feel they are doing enough. Just one example of this is the fact that they never have enough money or space to hire the number of speech therapists necessary, resulting in longer wait times for an appointment for pediatric speech therapy. "It's the poorest children that don't have a lot of options. And that's not just Shreveport, that's everywhere," says Sudduth.

The five women clustered around the file of the first child of the day, reviewing his history, planning who would do what. They work in teams so that they can see two patients at a time. They chatted and prepared a little play area for the child on the floor of the pediatric therapy gym. Katie Hebert, a young occupational therapist, dragged several of the colorful padded play mats and a clear plastic bin of toys into the center of the room. Along the sides of the room were various braces and small walkers, swings, activity tables, and toddler seats. One wall was painted with a colorful mural of a beach with a smiling sun in the sky.

Hebert grew up in the area and has worked at the clinic for three years; she loves pediatrics because she feels that she is supporting the whole family during what can be a difficult and confusing time. "If there are any red flags, we pick up the kid for weekly therapy," she explained. "Or if they live too far away to come here weekly, or the financial situation is too much of a strain, we can refer to Early Steps [Louisiana's Early Intervention program], where a therapist can come to the home. And a lot of times we are the first set of hands and eyes that notice any neurological red flags, like abnormal muscle tone or abnormal reflexes. Then we communicate that to the doctors and they order a neuro consult."

Generally, physical therapists deal with gross motor skills— crawling, standing, jumping, climbing stairs—while occupational therapists handle the fine motor skills, like holding a crayon and using a spoon. But in practice, with very young kids, occupational and physical therapy assessments can blur into each other, and both

kinds of specialist at the Willis-Knighton clinic use the Early Intervention Developmental Profile, a screening test that gives an
overview of a child's developmental progress. Are they making eye
contact? Holding up their head? Grasping toys? Pulling up to stand?
The speech pathologists use the Rossetti Infant-Toddler Language
Scale, an assessment that evaluates a child's preverbal and verbal
skills from infanthood through three years old. Communication
between the baby and caregiver is, or should be, developing even
before the first words are spoken.

The first baby arrived around 8:15, slung on his mom's hip.
He was twelve months old and had been born at 35 weeks; he
was blond and pudgy, his mom blond and slight. She had her hair
piled on her head, wearing ripped jeans and a purple T-shirt. The
romper-clad baby was babbling, excited about life, with a mouth
that naturally curved into a large grin. The therapists gathered
round and made oohing noises at him. Hebert, the occupational
therapist; Kari Johnson-Torres, a physical therapist; and Julia Bryn,
a speech pathologist, slipped their shoes off and joined mom and
baby on the mat.

Someone handed the child a ball. "BA!" he said, shoving it into
his mouth delightedly. The three therapists reached for him in swift
and coordinated turn, finishing each other's sentences, working
through their checklists, asking questions, watching the boy carefully, kindly.

"How's he been since we saw him last?" asked Johnson-Torres.

"Really well," said his mother.

"No illnesses or hospitalizations?" Johnson-Torres confirmed,
and the mom nodded.

"Oh my goodness, are you walking by yourself?" Johnson-Torres
asked the boy.

"Not yet. He's almost got it," says his mother, corralling the boy and patting his diapered bum affectionately.

Bryn tossed the ball to the boy, and he batted it down. "Good catch!" Bryn enthused. "You want the *ball*?" she asked him, taking it back and making an exaggerated show of the word. "Is he putting any sounds together?" she asked. "Like 'Mama'?"

"Just babbling," the mom reported. Bryn pressed a button on an activity table and music started playing. The baby made an exclamation of delight and pulled himself up to stand holding the table. Everyone clapped and said, "Yaaaaay!"

The questions came gently, conversational but quick; everyone knows a twelve-month-old's cooperation is a tenuous thing. Is he playing peek-a-boo? Can he hold a crayon? Bryn held him on her lap with a plastic mug and spoon held in her hands in front of him. "Yum, yum. Can you stir it?" She wanted to see if he could grasp a spoon and also if he knew what a spoon is for. Hebert asked if she could give him a snack puff, and the boy reached for it in the tipsy way of toddlers. He couldn't quite hold one at a time with his thumb and pointer finger, a pincer grip, and Hebert was slightly concerned about it, assessing that he had a bit of a fine motor delay. She explained to his mom how to encourage him to pick the puffs up one at a time with his fingers.

"Have you heard of Early Steps?" Hebert ventured. The mom shook her head no. "They come to your home for therapy. That will be more convenient for you than the drive here. Let's get that going and then we'll see him back here, since you travel far. Does that sound good to you?"

"That's fine, that's fine," said the mom, accepting the referral from Hebert as Bryn walked around cleaning up the toys they'd used, singing *"Clean up, clean up, everybody, everywhere,"* under her breath. The little boy screeched at her happily.

The pace of the day picked up: a parade of babies and toddlers two at a time, a mix of races and ethnicities, along with their moms (and it was all moms). The therapists worked with concentration in the bright room, each picking up from the other. With each child they smiled hugely, to be friendly and reassuring, yes, but also to see: How did the child respond? Was the baby making eye contact? Was she tracking faces? The women put their faces close to each baby and made soft *choo-choo-choo* noises, burbled their lips. "Hel-l-l-o-o-o," they said. "Look at you, sweet girl; look at you, sweet boy." They moved their faces left and right, making sure the baby followed, or shook purple and green Mardi Gras beads to each side to see if the babies could turn their heads. For the most part, the babies obliged, seeming totally charmed.

The tasks children need to do become more complex as they grow. So what at first seems like a minor issue can snowball into a more serious one without therapy. A baby with a minor, almost imperceptible motor delay can grow into a child who can't hold a pencil or run or climb stairs without falling. To a teacher, a child might seem oppositional, but a clinician who understands prematurity—who, in Als's terminology, is listening to what the child is really saying and asking for—might be able to identify a sensory issue that is preventing the child from paying attention. These therapists aren't holding the kids to an arbitrary typical standard; they aren't trying to "fix" them. Instead, they are trying to help these children find their own developmental cascade: how they can feel strong and confident; how they can be at ease. It means listening to premature kids as they grow, understanding their body language, guiding them toward the healthiest ways of developing and learning.

That's why the minutiae of these assessments are so incredibly crucial to a child's later functioning, and therefore their sense of themselves. "If kids fall through the cracks and aren't noticed until kindergarten, they may exhibit things like inattention and impulsivity, coordination difficulties, and muscle weakness," Hebert told me. "This child is then at risk for poor self-esteem, which lays the foundation for his or her life."

Later in the morning a very tiny two-month-old baby boy came in, cradled in a car seat carried by his mom, a blond woman wearing a lacy white blouse. Hebert took them into her office, off to one side of the gym, and spread a clean blanket over a mat on the floor. The baby had a sweet button nose and scrawny pink legs and feet poking out of a blue gown embroidered with gold giraffes. He had been born at 34 weeks weighing 3 pounds, 5 ounces and with neonatal abstinence syndrome, meaning he went through withdrawal from opioids his biological mom had used during pregnancy. His adoptive mom placed him on the blanket to change his diaper. She was a single pediatric nurse who lived in Shreveport. She said she had been waiting for a baby for twenty years. "He's my only. My first and only," she told Hebert. "He *was* itty-bitty. He's getting chunk-a-chunk now!" (Chunk-a-chunk is relative.)

Hebert slipped off her shoes and sat next to the baby on the blanket. She spoke softly to him, studied his face. He gazed at her and around the room with his newborn gray eyes and then bent his chickeny legs up to his chest and burbled. Hebert leaned against the wall, picked up the baby, and placed him on her legs. She ran her hands over the soft spots on the top of his head.

"Hey, buddy!" she said softly. "Do you think he is pretty calm?" she asked.

"Oh yes, he's very chill," said his mom eagerly.

Hebert asked about eating and sleeping as she examined the baby, explaining that this particular boy was at a higher risk of developmental delays and attention and mood problems because he was both premature and had been exposed to drugs prenatally, a not-uncommon pairing.

His mom was mildly defensive on his behalf. "One of the pediatricians did tell me he may be developmentally delayed," she said. "And of course the mama bear in me was just, like . . ." She sniffed loudly in a theatrical show of dismissiveness. "And he"—she gestured to the baby proudly—"turned around and rolled on his side and looked at the doctor. And the doctor said, 'Well maybe not!' My son is smart. There is nothing wrong with him."

Hebert asked if the baby brought his hands to his mouth, and the mom nodded. "And I read with him and I sing to him every night and I try to make sure he can follow his toys. I've been trying to read everything as far as his milestones and make sure he's doing what I think he's supposed to be doing."

"He looks great so far," Hebert said. "Just looking at his movements." She checked the baby's reflexes, laying him on his back and then holding his arms to pull him up to a sitting position. The baby wobbled but tucked his chin, helping to hoist the weight of his own bald head. "Good," Hebert whispered. "That's just what I want to see." She checked that he picked one foot up when held in a standing position on the floor, the stepping reflex. "Does he tolerate tummy time?" His mom affirmed this, so Hebert put the baby on his tummy and he started to fuss. "Oh, nugget, okay," said Hebert, rubbing his back. Both adults cheered when he haltingly lifted his head up.

At the end of the assessment, Hebert assured the baby's mom that he looked great. "He's definitely age appropriate for his gross and fine motor skills, at least for his adjusted age [almost one month], but he's really closer to the two-month mark. He looks really good. He's precious and you're doing a great job." They agree that she will see him again in two months.

This particular visit made me think about how developmental care is for the parents as well as the child. (This circles back to Als's emphasis on the baby as part of a family and the wisdom of treating the family holistically.) Even though this mom was a pediatric nurse and clearly already knew a lot, she seemed hungry for Hebert's expert assessment, for her reassurance and for information about what to look for next, strategies to help her son develop healthily. I saw this to a greater or lesser degree in every parent who came to the clinic that day—a mixture of pride and worry, a desire for reassurance and information. Even when a delay was diagnosed, it often seemed a relief, because treatment was also offered.

Also, being among therapists who understand premature kids so well can feel like a reprieve, because sometimes it seems like others—the moms in your mom group, your neighbors, your day care providers—don't understand where you've been with your child. Here, everyone understands. They know why you're proud and they know why you're worried.

A tired-looking mom toted in her six-month baby boy, wearing plaid Madras shorts and a T-shirt. Johnson-Torres sat cross-legged on a padded platform and took the baby onto her lap. He had been born at 34 weeks and suffered a small bleed in his brain, which put him at higher risk of developmental delays. He also had to have a portion of his intestines removed two days after birth and had an ostomy bag for three months, which was then reversed in a second operation so that he could digest food normally. His mom's eyes looked tired. She sat straight in her chair and answered Johnson-Torres's questions in quiet detail.

Johnson-Torres held the boy in her lap and asked how he had been since he was last assessed. "We saw him in May; I think it was right after the surgery," said Johnson-Torres. "So no hospitalizations since then? I'm just going to look at his scar if you don't mind." The mom assented, and Johnson-Torres lifted up the baby's T-shirt to examine his tummy. Then she checked his reflexes, flexing his feet and gently pulling him up to a sitting position. She laid him back down and dangled green beads over his head to try to get him to follow them in both directions. The baby kick-kick-kicked his legs and reached for the beads.

"Kicking those legs, bud? You want the beads?" The baby followed the beads to one side but not as much to the other. "You do have a preference, don't you?" Johnson-Torres leaned her face close to his and massaged the side of his neck. The baby fussed. "Yes, yes, you do. This side is tight. You feel it, don't you?"

"Can you see, Mom," Johnson-Torres said, "how his chin goes a little bit further when it rotates this way than this way? You probably already knew this." Johnson-Torres was still massaging the baby's neck and the baby started to cry. "I know, I know. I want to feel it," she said to him. "Got some tightness there." She started singing to him, "Twinkle, twinkle little star . . . ," and pulled him back up to a sitting position.

Johnson-Torres addressed the mom again: "Does he do okay with tummy time?"

"He likes it some, but he fusses," his mom said. "We were behind on tummy time because of the stitches. He was on his back so much."

"Given the circumstances, it's expected," Johnson-Torres reassured her. "So, Mom, I would like to pick him up for PT. It's going to just be some stretching to start. A lot of it's going to be education, things for you to work on at home. We want to make sure skills come along and that muscle gets stretched out. It's better to address it now, while he's young. This muscle"—Johnson-Torres tilted her head and gestured to the muscle running along the side of her neck and shoulder—"attaches to the jawbone. It can alter the jawline, ear placement, eye placement, so we want to be on top of it."

The mom nodded, and Johnson-Torres started to show her some gentle stretches she could do at home with her son. She asked if the mom had any concerns.

"He takes the bottle well, but he's pulling away from food," she said, describing how she had started offering him solids three weeks earlier but he didn't seem to like it, turning his head away from the spoon. Johnson-Torres listened and then waved speech pathologist Bryn over; she also has expertise in feeding issues, which are common for preemies.

Bryn listened to the mom repeat the story and then asked if the boy is gaining weight. "Well, he had part of his intestines removed; I think that's why he's not *fat* like most babies his age," said his mom.

Bryn said that she thought it was a matter of exposure: Eating is a new experience and it would probably get easier for him. "If you want, put his hand in the food, help him feel it, let him see it in the bowl. First we have to see, then smell, touch, and that tells us if it's okay to eat. Let him play with it and get messy. I know it's not easy."

"Oh, no, I don't mind," said the mom.

"You know how some parents scrape the food off the baby's face with the spoon?" Bryn asked. "No, let it stay there. Encourage him to use his tongue to feel that it's there, to realize he needs to lick or wipe it off. Let him get dirty. You know, it's all one big system," she said, referencing his intestinal surgeries.

The mom and Johnson-Torres regrouped and decided to meet again in a month to start physical therapy and to give the boy some time to adjust to eating.

Meanwhile, in the opposite corner of the gym, Hebert was cuddling a three-month-old baby boy who was looking around with big, soft brown eyes. His mom, a chatty woman with olive skin and long, wavy brown hair, told me that her oldest son was born at 27 weeks, then she carried her middle son to term before giving birth to this one, her youngest, at 30 weeks.

The baby's pediatrician, she said, had diagnosed the baby with laryngomalacia, a condition in which the airway is floppy instead of rigid, as it should be. This can manifest in breathing difficulty, as the airway folds in on itself. But Hebert wasn't sure that the baby was exhibiting the signs of this disorder—although she was careful to say that she couldn't diagnose it or rule it out because she wasn't a physician. Bryn came over to take a look and said that, in her experience, babies with laryngomalacia usually have stridor, a high-pitched wheeze with each breath. She didn't hear stridor from this baby at the moment.

Hebert and Bryn looked at each other and agreed that the baby needed a second opinion from an ear, nose, and throat specialist. Hebert proceeded with the assessment, the asking and answering and soft baby talk that had been the rhythm of the day. She noticed that this baby wouldn't tuck his chin when pulled up to sit, an instinct that is part of learning how to support your own head.

"His skills are looking pretty good, but I would like to see more head control," Hebert said. The baby looked at her and gurgled happily. "Yes, I'm talking about you!" she said. "It could be because of structural issues," she pointed out, referring to the possibility of a floppy airway. "We need to look into things further, so I'm going to refer you to an ENT."

"Just give me the address and I'll be there," said his mom.

Hebert showed the mom an exercise to do in the meantime, to get the baby to practice tucking his chin. She leaned back and held the baby above her face, so that he had to lower his chin to look into her eyes. She clucked and cooed at him, tried to get his attention. "You gonna spit up all over my face, huh? There!" she said as the baby wobblingly obliged and lowered his head to gaze at her. "See how he's trying? We definitely want to see him again soon to see if that improves, so we know if there's a reason he's avoiding that."

It takes more than a village to raise a premature baby. It takes a village and a university and a hospital and a pediatric rehab center. It takes caring and resources of every kind, often for years.

Kesha Evans, a Washington, D.C., lawyer, summarized this in a way that felt really true to me. We were talking about how premature birth changes your perspective—how you have to adjust your ideas about what being a mother will be like. When she was pregnant, she had carefully laid out her intentions: She had a birth plan and clear ideas about how she and her partner would raise their son. But when he was born unexpectedly at 27 weeks, spent five months in the NICU, and then came home with complex medical needs, all her plans evaporated. Her perspective changed dramatically. She realized that there was a necessary collectivism to caring for a premature baby.

"It was me, the mom, with the stroller, with the feeding tube attached to it, and sometimes the cannula, walking down the street. Realizing that it was his nurse, and his physical therapist, and his occupational therapist, and his doctors, and his cardiologist that were as much a part of this as I was. And his grandmothers. Without all of these people, he would not be here, manifested the way he is. The community that it took was perspective shifting. So that broader perspective of parenting was my takeaway. And that, I think, is a gift."

The trajectory of neonatology—being able to save ever-smaller babies with technological advances—is extraordinary, but it is only the first step in these children's lives. Maybe we'll be able to grow babies in bags in the near future, but we'll also need people to do the relatively unglamorous, underappreciated work of making sure that these children can turn their heads to both sides as they grow, that they aren't being held back by a subtle sensory issue, that they can pick up a snack and hold a crayon and make eye contact and know when someone is smiling at them. And if they struggle with some of those things, we need professionals who will help them adapt and professionals who will adapt their environment to them.

In a way, parents of preemies just come to this realization sooner than most: You can't control what hardships your child will encounter. You can't prevent them from experiencing pain. You can only do your best for them. And none of us can do it alone.

Part V

The Threshold

End-of-Life Issues at Birth

12

What Should We Do for 22-Week Babies?

FOR A VERY long time, 24 represented a kind of magic number. Starting in the 1980s, 24 weeks' gestation was thought of as the moment when babies were reasonable candidates for resuscitation in the delivery room and the battery of NICU supports. Without those active steps, babies born that early inevitably die within minutes or hours. But this line—which is not really a line at all but a zone, a hope, a question—has crept lower, and now 23-week babies are often treated. Less often, 22-week babies are, too. Whether or not that is a good thing depends very much on whom you ask, how you slice the statistics, and where you are—what hospital, what country.

This dilemma is also tangled up in factors that are even harder to pin down: feelings and values and projections about babies, suffering and disability, life and death, parental authority versus medical expertise. In the United States, there is perhaps no institution more committed to the rightness, the justness, the appropriateness of attempting to treat 22-week babies than the University of Iowa.

University of Iowa Stead Family Children's Hospital is in Iowa City. It is smack-dab in the middle of the Midwest, a bucolic college town where almost half of all residents are employed by the university. It is charmingly sleepy, cerebral, unfailingly polite.

If you were to drive due north from Iowa City, through Cedar Rapids and swaths of starkly beautiful farmland, fallow black and ocher in winter, about sixty-five miles later you'd find yourself in Manchester, Iowa.

Downtown Manchester consists of a few blocks lined with shops like Bushel & a Peck, where the town's 5,000 residents can buy a dozen different kinds of homemade sausages, baked goods, and fruit preserves. There's Widner Drug, a consignment shop, Aunt Emmy's Bakery, several cafés, a quilting supply store, a Chinese restaurant, a pub, a Hardee's, and a Walmart. The smaller streets that grid out from the center are quiet and well-kept, lined with middle-class homes, churches, schools, two playgrounds, and a skate park.

Down one of those residential streets sits a neat white gabled house with a wide front porch and a couple of trucks in the driveway.

Inside that house, on a sunny afternoon in February 2018, Alexis Hutchinson was flying. The seven-year-old had a blanket tied around her shoulders, and she raised her arms to make the fabric billow out behind her as she whooshed around the living room with her sister Kinsley, age three. Alexis was grinning, her sweet face framed by a dark-blond bob. The girls were pretending to be Herky the Hawk, the University of Iowa football mascot. They bounced off the furniture and each other, resisting all parental efforts to corral them. Max, the family's elderly black-and-white chihuahua, hopped around their feet, trying to keep up.

Their older sister, Joslyn, twelve, was organizing Girl Scout cookie orders. Isaac, their older brother, was upstairs. (He was fourteen; who could blame him?) Their mom and dad, Chrissy and Jordan, were sitting on the couch, fielding requests for pancakes and TV shows and talking to me about Alexis's extraordinary birth.

Chrissy and Jordan both grew up in Manchester and knew each

other in school, although they didn't start dating until after high school graduation. Chrissy became a pharmacy technician, and Jordan started working at Exide Technologies, a company that makes car batteries and is one of the biggest employers in the area. In 2003, Chrissy gave birth to Isaac, and Joslyn followed in 2005, both full-term, healthy babies.

In 2010, when she was twenty-one weeks and five days pregnant with her third baby, Chrissy's water broke in the middle of the night. It wasn't a gush so much as a slow trickle; she thought—she hoped—it might be pee. She and Jordan went to the closest ER, where it was confirmed that she was leaking amniotic fluid. They had no experience with prematurity, no reference points for this kind of calamity. They didn't have a particular religious faith to guide them. They just wanted their baby alive and healthy.

They drove to St. Luke's Hospital in Cedar Rapids, where doctors said all they could do was to try to keep the baby in the womb as long as possible. But if Chrissy got an infection—a risk when membranes break but labor doesn't start—delivery would be inevitable. And at 22 weeks, doctors explained, they would not offer NICU treatment, only comfort care. In this context, comfort care means wrapping the baby in a blanket and having her parents hold her until she dies. It was the hospital's policy, and not an uncommon one.

Two days later, Chrissy hit 22 weeks, but that night she started running a fever, a sign of infection. The baby had to come out. Doctors started the process of inducing labor, and Chrissy and Jordan started trying to wrap their minds around the idea that their child was going to be born and then they would have to watch her die. Jordan remembers asking three different doctors at St. Luke's if they would just try to treat the baby and being turned down. (A spokesperson at St. Luke's confirmed that this was their policy in 2010 but said that the hospital now allows resuscitation at 22 weeks.)

A family member suggested talking to someone in Iowa City to see what their options were. Chrissy remembers doctors in Cedar Rapids saying, "You can call, but University of Iowa's going to tell you the same thing."

"And then we called the doctor in Iowa City," Jordan remembers. "And he said, 'We'll do as much as we can to keep the baby alive.'"

"I was really scared and I didn't know what would happen," Chrissy said. "We thought, you know, the baby deserves a chance."

So, as Chrissy went into labor, she got into an ambulance and traveled the twenty-six miles to Iowa City, to the Stead Family Children's Hospital. They arrived around midnight on April 25, which made Alexis exactly 22 weeks and one day when she was born at 9:10 a.m., weighing 1 pound, 1 ounce.

Dr. Jonathan Klein was the attending neonatologist in the delivery room that morning. He explained that the hospital had uniquely good results with extremely preterm babies but that he couldn't guarantee anything. Chrissy had not been offered steroids at St. Luke's—if the baby is not going to be resuscitated, there's no reason to give steroids—so Alexis lacked that advantage. Klein explained that he would intubate the baby and try to get her heart rate up and her blood oxygenated, but he wouldn't go as far as chest compressions. Chest compressions are a last resort in neonatal resuscitation, and the hospital generally doesn't do them on 22-week babies.

"So she was born," Klein told me later. It was a vaginal birth, because, generally, at this gestation, C-sections are too risky to the mother when weighed against the baby's chance of survival. "And she had a heart rate less than 60. And she was so translucent I could see her heart beating. Like this . . ." He mimed with a fist bumping against his other palm. "So we intubated and bagged, and I could *see* the heart rate just speed up."

Normally, clinicians in the delivery room take a baby's pulse by feeling the umbilical artery while waiting for the monitors to kick in, which usually takes a little over a minute. In this case Klein could actually watch Alexis's heart rate increasing as she responded to resuscitation. "I said, 'Okay, she wants to live,'" he said.

For Jordan and Chrissy, the birth was a blur of intensely conflicting emotions, hard to take in. "When she was born, you could see right through her skin and everything," Jordan remembered. "Her eyes weren't even open. Her ears were still connected to her head."

"They took her over to the table right away, and they had the bag on her. It didn't take them long, and she was breathing," Chrissy remembered of the moments right after Alexis's birth. "Jordan went over there and put his finger in her hand, and she did squeeze it. Then, before they took her to the NICU, they brought her over, and I kissed her on the head before they took her out. I just couldn't believe how tiny she was. It was crazy."

Chrissy has long, straight light brown hair and dark eyes; on this particular day almost eight years later, she was wearing a relaxed T-shirt and black stretch pants. Family photos lined the walls behind her. She was calm and thoughtful as she described the morning of her third child's birth, but sometimes she just trailed off and went quiet when I asked an unanswerable question—"What were your thoughts when you first saw her?"—and the inexplicableness of it all was palpable in her silence.

And here was Alexis, nicknamed Lexi, bouncing around the room, wearing a beaded necklace and a shirt that read "Brave, Fierce, Kind." She brought out a small photo album for me to look at. Inside were photos of her throughout her NICU stay; the photos themselves seemed an expression of hope, a grasp at normalcy. Here was baby Alexis, one pound, with translucent red skin, lying under plastic. Here was a photo of a diaper on a scale, a funny detail that I remember, too, because although your impulse is to throw a diaper out, in the NICU, all fluids are quantified and recorded.

Alexis spent her first 145 days in the NICU and responded remarkably well to treatment; she never had a brain bleed, never got an infection, never got necrotizing enterocolitis, escaped all the dire complications that stalk babies like her. She went home on oxygen, which she needed for about a year, and a feeding tube, which she needed for about five months. She has been hospitalized a few times for strep throat and pneumonia—a common occurrence for preterm kids, who can have less effective immune systems—but is otherwise entirely healthy. She likes to read, and she wants to be a veterinarian when she grows up.

Chrissy and Jordan know that things could have gone differently. They know that if they hadn't gotten themselves to the University of Iowa, Alexis wouldn't be here. They know that, even at Iowa, Alexis might not have survived, or might have survived with profound, life-altering disability.

Now they focus on their luck, on their happy child, but the experience offered plenty of trauma to go around. There was the baby they couldn't hold, the long months spent in the NICU, helpless, not knowing if she would survive. There were the two other children to care for; Isaac went to live with his aunt, Chrissy's sister, so he could finish the school year. There were Jordan's endless drives back and forth from home to work to the hospital, since he had to go back to work or jeopardize their health insurance. When they finally brought baby Alexis home, Chrissy and Jordan had to change her feeding tube every week, threading it up her nose and down her esophagus, checking to make sure that they hadn't accidentally run it into her lungs instead of her stomach.

But what really gets to them, even now, is the idea that at St. Luke's the decision was not theirs to make, while at the same time, twenty-six miles away, their baby, the very same patient, was offered the choice of comfort care *or* intensive care.

"I think, you know, with any preterm baby, I think it could go either way when they're that little. You don't know what they might have, complications-wise," Chrissy said. "I just feel like they deserve a chance, and [that we] try."

I asked her what she would tell other parents in her predicament. "I would say do what you think is best for you and your baby and family," she said. "I don't know. For me, I wasn't done, you know. To give her up or not fight for her. It's hard. I know there are people who are, like, 'No.' I guess they don't want the baby to go through all that."

Jordan interjected, "The baby and you, they [the doctors] don't want you to do all this and then all of a sudden something bad could happen."

"That's the risk," said Chrissy.

When Chrissy and Jordan say that people don't want babies to go through "all that" because "something bad" might happen anyway, they are using a shorthand for the almost unimaginable intensity of NICU care for a 22-week baby—the endotracheal tubes, the intravenous lines, the blood draws and transfusions, the medications, endured by a child who barely has skin—balanced against the constant, real threat of death.

Alexis's excellent outcome is unusual, at least based on national statistics, although obviously it is not impossible. Then again, anecdotes like this one are not data, and happy-ending stories of perfect "miracle babies" can give parents cruel, false hope and create a narrative that's hurtful for those who weren't so lucky. If Alexis did have disabilities—and there are many others like her who do—it would not diminish the value of her life. Alexis is a person, not an argument. She doesn't have to be a miracle baby; she is enough just as she is.

Nevertheless, her life does make a point, and it's one her parents feel comfortable making: Children like her can survive and go on to be happy and beloved. What does it mean that some hospitals will not offer active treatment for 22-week babies?

Life and death are never so closely entwined as when a baby is born far too early. There are only about 5,000 babies born at 22 and 23 weeks' gestation in the United States every year, but they are the subject of much debate because there is no consensus on whether to actively treat them—attempt to resuscitate them, put them on a ventilator, bring them to the NICU—or give them comfort care—let them die—in the delivery room.

It is rare that a family is able to take the kind of initiative that the Hutchinsons did, switching hospitals in the midst of the crisis. For them, it was a fortunate confluence of being certain about what they wanted, being close to a hospital that would honor their wish, and having enough time to get there. For most families in that position, the hospital policy or physician attitude will shape their lives forever: for better, worse, or somewhere in between.

When there is conflict between parents and hospitals it can come in many iterations.

Lametria Burks, for instance, worked within hospital policy to get the treatment she wanted. Her water broke at 22 weeks exactly, and doctors at Madigan Army Medical Center, the hospital on the military base just outside Lakewood, Washington, where she lived with her husband and three older daughters, would not resuscitate until the pregnancy reached 23 weeks. They also wouldn't give her steroids until she reached that point. But then Burks realized that the policy was actually 23 weeks *or* 500 grams, so she insisted that a growth scan be done, which estimated her fetus at 600 grams, big for a 22-week baby. Burks used this new information to argue that she should get the steroids and that her daughter should be resuscitated at birth, and finally the doctors agreed. "I fought for my child," Burks said. "They were going to give my baby a chance." Her daughter, Carli, responded well to treatment and is now a chubby toddler who loves playing with her sisters.

The first question most people ask when they hear all this is: What *are* the survival rates for babies born at 22 weeks? Unfortunately, it is very hard to get a straight answer, first of all because the outcomes deeply depend on lots of factors, including characteristics of the baby like birth weight and sex, whether they got steroids before birth, and the reasons for the early delivery. (If there's an underlying health problem, like growth restriction or an infection, survival rates are lower.)

But getting an answer is difficult not just because of differences among babies. It's also because there are big differences among hospitals: Some hospitals have a zero percent survival rate for 22-week babies; others have 15, 20, or even 40 percent survival.

A few years ago, Dr. Matthew Rysavy, a neonatology fellow at University of Iowa, was trying to solve this puzzle. He was looking at outcomes for babies born at 22 to 26 weeks. Oddly, the survival rates at some hospitals were five times better than at others. All were high-resource academic centers; the care available in each should have been roughly the same.

At first he tried to look at the treatments used at each hospital to see if they made the difference. But that didn't explain most of the variation. When he dug deeper, he found the starkest differences in outcomes were for births at 22 and 23 weeks.

The short answer was that some of the hospitals were trying to treat those 22- and 23-week babies, and other hospitals were not; some were initiating potentially lifesaving care, and others were not. For the tiniest babies, the variation seemed to have more to do with the policies and choices being made than it did with anything else.

Rysavy didn't set out to study 22-week babies or to make a point about them. But when clinicians read his paper, it was hard to miss the deep implications. Essentially, it was the story of Alexis at St. Luke's versus Alexis at University of Iowa, but as a statistical analysis, and writ large across the population. It was a kind of lottery—albeit one that providers didn't mean to set up and that parents didn't know they were participating in.

Not only that, but because the common convention is to average the numbers from the hospitals, the statistics being given to parents were a bit misleading. The zeros were bringing the average survival way down.

Rysavy separated out the numbers for babies whom doctors had tried to treat from those who had been allowed to die. For all 22-week babies averaged together, survival was about 5 percent—and that was the number that many clinicians were using. However, looking only at babies that physicians *tried* to save—which is the more relevant information for someone who is wondering if their baby *can* be saved—the survival jumped up to about 23 percent, or, in a newer large multicenter study, 38.5 percent. About half of those survivors have a serious disability, a broad category that includes people who are, say, legally blind, people who use braces to walk, and people who depend on a ventilator to breathe and are unable to speak or walk.

Those numbers are not very encouraging, but they are very different from 5 percent.

If you are in labor at 22 weeks and your doctor says that your baby will have a 5 percent chance of survival without making it clear that the majority of those babies never got any potentially life-saving treatment, then you, as a parent, might be less likely to push for treatment. In other words, citing unencouraging statistics, in turn, creates more unencouraging statistics.

After Rysavy's paper, which was published in the *New England Journal of Medicine* in 2015, some hospitals reevaluated their guidelines, allowing for more deference to parents if they asked for treatment for their 22- and 23-week babies. At the very least, there is now an increased effort to make it clear, when talking about survival, whether or not the numbers refer to all babies or just babies who got treatment. But practices and beliefs are not easy to change. Certainly there are still hospitals that will not offer active treatment for 22-week babies under any circumstances. And almost no one outright *recommends* it.

The variability that Rysavy described within the United States happens among other industrialized nations, and it is just as jarring. In the Netherlands, babies under 24 weeks are not resuscitated; at 22 and 23 weeks, survival is zero. In Sweden and Japan, nearly all 22-week babies are resuscitated, leading to a 46 percent survival rate in Japan and at one center, in Uppsala, Sweden, a survival rate of 53 percent.

These are all rich countries, with access to the same information, education, and equipment. The babies' conditions are largely the same. But those nationalities are clearly not in agreement about what is *right* to do for extremely premature babies.

It is possible that hospitals (and countries) that routinely treat

22- and 23-week babies are much better at it, and have better outcomes, than hospitals that do it only occasionally. In other words, it may not be effective to concede to parents' entreaty to try to treat their borderline-viable baby every now and again; in order to have good outcomes, you might have to proceed in the belief that 22-week babies can be good candidates for treatment.

This is what the University of Iowa does. And the result seems to be the best survival rate in the country: According to the hospital's data—which, as I write this, has been accepted for publication in the *Journal of Pediatrics* at the end of 2019—an astonishing 58 percent of all live-born 22-week babies and 80 percent of all live-born 23-week babies survive. (These results are extraordinary, but it's important to note that the total number of babies we are talking about here is small: 24 babies born at 22 weeks and 52 babies born at 23 weeks, over a nine-year period. This is because it is quite rare to be born that early.) Not only that, but among survivors, only about 18 percent of 22-week survivors and 9 percent of 23-week survivors have a severe disability.

That makes Iowa City, arguably, the white-hot center of the neonatology universe: If you were in labor at 22 weeks, and you wanted your baby to get NICU treatment, probably the three best places in the world to be are Iowa City, Sweden, and Japan.

To figure out what Iowa is doing right, I went to see Dr. Edward Bell, the neonatologist responsible for shaping the policy there. Bell, who was born in 1948, has a distinct Mr. Rogers–meets–George H. W. Bush vibe. He's tall, clean-shaven, tucked in, soft-spoken. He grew up on the western edge of Appalachia, in Washington, Pennsylvania, a small steel town. His father was a physics professor at nearby Washington & Jefferson College, and his mom, who recently died at the age of one hundred, was a stay-at-home parent.

Bell was a conscientious objector to the Vietnam War. He went to Columbia University for medical school. There he found he loved the neonatal unit, especially the adrenaline rush of the delivery room resuscitations. "Knowing what to do and how to do it technically, you could, in those first minutes of a baby's life, have an impact that would affect their whole lifetime," he said, remembering what attracted him to what was then the new field of neonatology.

One of those days, as an intern in the delivery room in the mid-1970s, he made a mistake. At the time, babies under 1,000 grams (about 2.2 pounds) were not considered viable, and the policy was not to resuscitate them. "I forgot to weigh the baby," Bell remembered. "I intubated it and it weighed 960 grams."

It would be convenient if this story showed that Bell was already questioning restrictive policies. In that version of events, of course, the baby would do well and young Bell would be vindicated. That's not how it was, though. At the time, the 1,000-gram rule didn't bother him. "I didn't think to question it," he said. "It wasn't because I was a rebel. I just forgot to weigh him." As for the baby: "I don't know if he ultimately survived. He had a pretty rough course. Had bad lung disease and he had a brain bleed and hydrocephalus. The state of the art then was not very good. There wasn't much we could do. We didn't even have ventilators for babies yet; we were using the adult ventilators."

It wasn't until 1976, during his fellowship at McMaster University in Ontario, Canada, that Bell started wondering about how those life-and-death decisions were made for the smallest babies at the moment of birth. He was studying under Dr. Jack Sinclair, a prominent early neonatologist. One day Bell was on service with Sinclair, and a 300-gram (about 10.6-ounce) baby was born. Sinclair asked

Bell to try to resuscitate that baby, even though the very smallest survivors they had at the time were at minimum 700 grams.

"Everyone was appalled that Jack encouraged us to resuscitate this baby and treat it. Everybody thought this was morally offensive," Bell told me. "But Jack said, 'How are we gonna learn if we don't try with some of these smaller babies? How are we gonna know what's possible?' I didn't share his thinking at the time but, looking back years later, I can appreciate what he was saying." As for that baby, he didn't live. But he lingered in Bell's mind. That question lingered, too: How are we going to know what is possible if we don't try?

By the time Bell landed in the University of Iowa NICU in 1979, the consensus was that if a baby's eyelids were still fused, they were too young to live; and if the eyes were open, they might be viable.

"The thought was that that tended to line up with the critical states of lung development," he remembered. "We now know that that's not true." The problem, of course, is that not every baby's development proceeds in such a neat, orderly manner. "Some babies' eyes would be open at 25 and some wouldn't open until 27," Bell said. Then they had a 26-week survivor whose eyelids were fused at birth. His name was Sam and he confirmed what Bell suspected: Eyelids weren't a great criterion for deciding to resuscitate in the delivery room or not. By the late 1980s and early 1990s, antenatal steroids and surfactant became standard, making it possible for younger and smaller babies to overcome lung immaturity and respiratory distress.

Bell started to notice a pattern: The practical viability line was moving and the criteria by which these decisions were made—the ones that everyone had been *so sure of*—kept blurring and changing.

Sometimes this progress was almost a matter of shopping. In

1994, obstetricians contacted Bell about a woman whose fetus was extremely growth restricted—27 weeks, well past viability at the time, but under one pound. The plan was to try to treat the baby when she was born, but Bell realized that the smallest endotracheal tubes the hospital had on hand would not be small enough for this baby's airway. Bell asked one of the nurse managers to order 2.0-millimeter endotracheal tubes—that's 0.07 inches in diameter—the smallest available. "She thought it would be ethically questionable to resuscitate a baby that was so small, it would need a 2.0[-millimeter] endotracheal tube," remembers Bell. "And she stood her ground and said, 'I'm not ordering those.'" So he tracked one down himself and carried it around until the baby was delivered; it was a girl weighing 359 grams (12.7 ounces), who survived. From then on, the hospital always stocked those tubes.

Bell likes to say, "If you don't push a wall, you won't know if it moves." So he recruited other doctors who were comfortable working this way and built a culture based on the notion that progress comes from pushing, from being willing to do things others aren't. (This reminds me of Delivoria-Papadopoulos and her insistence on trying, over and over and over.)

This, maybe, is the crux of the disagreement: Some doctors feel that a baby's life should never be a let's-try-it-and-see situation, especially given the potential suffering involved in NICU treatment, and the very real power imbalance between families and doctors. Doctors at Iowa feel that sometimes you *have* to proceed this way at the youngest gestations. They also feel that because their results are good, the burdens the babies bear have a reasonable chance of being balanced by benefits—the benefit of getting to go on and live a long life, most of all.

The first 22-week baby Bell treated was born around 2000. "I was called back to labor and delivery to talk to a woman who was going to deliver at 22 weeks," Bell remembered. "She asked specifically to talk to a neonatologist, though ordinarily women at that gestation weren't referred to a neonatologist because treating the baby was considered futile. They weren't admitted to the NICU. I told her that, and she said, 'Well, Dr. Bell, you told me the same thing five years ago about my 23-week baby, and she's doing fine, so I want you to try with the 22-week baby.' I could hardly say no." That baby, the one with a 23-week sister, died. But her mother was glad they had tried. And Bell's evolution was already under way. Later that same year, he had his first 22-week survivor.

Two decades later, Iowa's policy is to recommend NICU treatment for 22-week babies, although comfort care is also offered as an equally valid choice. Few parents opt for it, though, perhaps because once the idea of hope is raised, they jump at that chance, or feel that they must. At 23 weeks, comfort care is not offered, although if a parent really wants it, it's an option. At 24 weeks, resuscitation is mandatory. This is in contrast to many hospitals, which offer comfort care as an option through 24 weeks.

Dr. Brownsyne Tucker Edmonds, an obstetrician-gynecologist at Indiana University who researches decision-making in extreme prematurity, has noticed a generational divide, with older neonatologists sometimes more interventionist than younger ones. "If you think about the neonatologists who have been practicing for thirty years—if you think about the advancements that they've seen and the way the needle has moved on survival—they are a pioneering generation with a pioneering mind-set. . . . I don't blame them for thinking: 'Of course we should be pushing the envelope, because that's how we've made the progress that we've made to date.'"

Tucker Edmonds sees all sides: She agrees with the current

consensus, which is that *recommending* active treatment at 22 weeks is not supported by most of the published evidence. At the same time, she feels that no information should be kept from parents, and it is incontrovertible that some 22-week babies do survive. It is particularly sticky, she says, because once a physician brings up the possibility of active treatment at 22 weeks, it can become a kind of offer parents can't refuse. "If you [a doctor] make a claim that there's something that *could* be done, there are a lot of people who can't really kind of wrap their heads or their hearts around the notion of not trying." She points out that parents often focus on the possible: *You're saying it's* possible *that my 22-week baby might survive?* And the physicians' point of view is: *Yes, but that is not* probable. She sees a kind of pendulum, a push-pull between hope and realism, the individuals and the statistics, and she is trying to find the best ways to acknowledge both. "I think that we providers have a responsibility to help create a space for families to hold both of those things, the possible and the probable," she said.

Perhaps more than many other neonatologists, Bell seems to feel comfortable living on the possibility side of the pendulum; then again, his survival rates are so uniquely high that he can also make some claim to the probable. Still, he's fully aware of the sorrow and struggle that are intertwined with what he does. He is haunted, in particular, by one girl who was born at 25 weeks about twenty-five years ago. She was left with severe cerebral palsy and developmental disabilities that made it impossible for her to live independently. "I told her mother that she would be able to walk," Bell recalled, "because that was fully my expectation. I've learned from that not to make guarantees. There's no explanation for why some babies have poor outcomes and some other babies who we think are going to do horribly turn out just fine."

Bell and everyone on his staff are believers in something that medical people call trial of therapy or, more dramatically, trial of life; it means that when the prognosis is very uncertain, you use treatment to see if treatment will work. Withholding of resuscitation (letting a baby die in the delivery room) and withdrawing of life support (letting a baby die after lifesaving care has been unsuccessfully attempted) are often conflated, and medical ethicists say they are both morally permissible. They both have the same worthy goal: the peaceful death of a child who won't benefit from intensive care.

But Bell sees a huge difference between them. If you decide not to resuscitate a baby at birth, you know what will happen: The baby will die. But if you try, you might get an Alexis Hutchinson.

Rysavy described trial of therapy to me as a betting analogy, an idea that originated in an *Acta Paediatrica* article by Dr. William Meadow, Dr. John Lantos, and several coauthors. If you're asked to bet on a baseball game before the game starts, you have *some* information—how the teams have done historically, current injuries, the weather forecast—but you don't have much specific information on which to base your bet. But with every inning that goes by, you've got a better idea of how things will turn out. Nothing's a sure thing until the game is over, of course. But if your team is up 5-0 in the seventh inning or vice versa, now you're much more confident about that bet.

Every day that an extremely preterm baby survives makes it more likely that they will continue to live. That might seem circular or obvious, but the problem is that you can't always tell which babies are which right away. A 2017 study showed that babies born at 22 weeks who are alive on day 7 have a 45 percent chance of survival: Surviving for one week more than doubled their chance of surviving overall. (This is also true for premature babies born at later gestations, but the effect is less dramatic.)

This is, perhaps, the most convincing argument for trial of therapy: It allows the survival of the maximum number of babies while allowing for the withdrawal of life support for those who would have died no matter what. It helps avoid what British bioethicist Dominic Wilkinson calls a "let die mistake," a horribly memorable way to put it.

Bell does not think that resuscitation should be mandatory for 22-week babies, and he wouldn't presume to tell other physicians how to counsel parents or run their units. He believes that parents should always have the option of comfort care at that gestation. But he *does* believe that writing off all 22-week infants as untreatable is both unjust and not supported by the data.

Dr. John Dagle, an attending neonatologist who works with Bell, was willing to put it more strongly.

"The equivalent of not resuscitating this baby is to pick up the phone and have the ER physician say, 'What? There's a car accident? The twelve-year-old is hurt pretty bad? Ah, just leave him there,'" he said, pausing for emphasis. "This whole trial of life is not anything crazy. To be honest, if a baby is previable, I can't save them. But I shouldn't decide who's previable. Plus, when you measure gestational age on ultrasound, it's plus or minus a couple days. So 23 and four is bad, but 24 and one is good? There's biologic variability! So the whole numbers game just bothers me."

I asked Dagle why he thought the University of Iowa has such uniquely good survival rates. In short, he said he feels that Iowa has good results with premature babies because Iowa values premature babies. "It has nothing to do with how smart you are. It's: Do you have the will? Do you value these kids? I think that's what a lot of this is about."

Dr. Jonathan Klein, the neonatologist who delivered Alexis, agrees that Iowa's try-it, believe-it culture is a big factor in their results. But he also thinks that Iowa delivers better care than most hospitals.

He ascribes some of the success to using high-frequency ventilation, in which tiny puffs of air are delivered to the lungs at many times the usual speed, instead of larger puffs of air at a slower rate. This way, oxygen is carried to the blood but the lungs never stretch out all the way, which can cause them injury, even rupture them. ("Like wet tissue paper" is how Bell described the lungs of extremely preterm babies.) Other hospitals certainly use high-frequency ventilation when they think it is appropriate, but Iowa uses it almost exclusively.

Klein believes there are other differences, too: "We keep our glucose levels much tighter, we keep our carbon dioxide levels much tighter, we keep our blood pressure much tighter," he told me. "We get a lot more labs. I'm not going to have a premature baby not get frequent labs the first week of life to save money for the hospital when half of all of the medical money is spent on the last six weeks of adults dying."

It's also crucial that obstetricians are in agreement about the approach. At Iowa, obstetricians will administer steroid shots at the end of the 21st week to women at risk of giving birth in the next few days. The American College of Obstetricians and Gynecologists (ACOG) does not recommend giving antenatal steroid shots until 23 weeks, citing lack of evidence at earlier gestations. However, ACOG says that it *is* permissible to resuscitate 22-week babies, a position that seems at odds with not giving antenatal steroids, because if you are going to resuscitate these infants, it seems unfair not to give them what they need to have the best chance. And according to a new study, the steroids more than doubled 22-week babies' chances of survival.

The act of pushing viability back to 22 weeks (and perhaps beyond) would be complex and contentious enough. Unfortunately, this is not just a medical or bioethical issue; it is also political, because of the direct link to abortion law via *Roe v. Wade*, which established that states may prohibit abortion after "viability."

Almost every neonatologist I talked to in Iowa took great pains to tell me that either they supported abortion rights personally or that they did not want abortion politics muddying the waters in neonatology. Perhaps this is because the news of Iowa's success has been co-opted by antiabortion activists.

In the context of abortion politics, it is easy to assume that doctors who are willing to resuscitate at younger and younger gestations are more religious or more conservative—more likely to be antiabortion. Actually, it's not at all clear that this is true. One study in Europe showed that the less religious the doctor, the more interventionist they tended to be in the context of extremely preterm birth. A study in the United States found a complex picture: Nonreligious doctors were more likely to offer pregnancy termination, but they were also more likely to perform a C-section for labor at early gestations, a more interventionist act that supports survival of the baby. In other words, they seemed more likely to offer more medical options overall. It's hard to make any big generalizations, but it is not necessarily the case that more religious or conservative doctors are more likely to be in favor of resuscitating 22-week babies, or vice versa.

Dr. Tarah Colaizy, an attending neonatologist at Iowa, sees her job as offering more parental choice, not less. "Ultimately, it's up to the parents," she said. "I'm very pro–extreme preemie, but I'm also very much pro–abortion rights. We're not—in fact, I would say that almost to a person—we are not staunch antiabortion people here. We're not."

I asked her if people sometimes assume that about her and her colleagues, and she nodded. Then she spoke about a little boy who was born at 22 weeks and successfully treated at Iowa. His parents bring him to testify against abortion to lawmakers. Colaizy is grateful that she was able to successfully treat the boy; glad that he is healthy and has done well. "But I don't like what happened there," she said.

Bell doesn't see a conflict between a preterm baby's right to care and a woman's right to an abortion.

"I keep them separate. The fetus is the fetus, and the baby is the baby. Pregnancy termination is legal," he said. "I'm not troubled by the fact that babies can be legally terminated at the same gestational age as I am treating them in the NICU. Some people find that unacceptable.

"Once they're born, they're humans and they're entitled to the same rights as other people, and their medical decisions should be made the same way as anyone else's," Bell said.

He is speaking within a legal and ethical tradition that sees birth as a fundamental transformation, making one body into two bodies. Abortion is medical care; so is resuscitation of 22-week babies. Medical care can involve complex ethical questions and difficult decisions. But those dilemmas, when they exist, are idiosyncratic; they are personal. Shared decision-making, the current ethical standard for healthcare, in part posits that every person should be able to apply their own values to the medical care they receive, and to shape their own lives.

There is difference of opinion, at the margins of viability, about which premature babies should get life-sustaining care. But no one argues, as they do in the case of abortion, that there ought to be a *law* delineating appropriate care by gestational age, which is not even a precise measurement. That would oversimplify terribly personal situations and would take the power out of the hands of patients (or parents) and doctors, who are the only ones with all the information.

These two medical treatments—abortion and attempted resuscitation of preemies—are only coupled because the decision in *Roe v. Wade* tied them together. The use of a changeable, debatable idea—viability—to define the right to an abortion has basically guaranteed the ongoing erosion of access to abortion. And it has needlessly complicated the ways that society views the treatment of premature babies.

Tying abortion access to viability, in a sense, pits premature babies, and their potential to live, against women and their right to access the reproductive care that is best for them. It is possible, on some level, that some people's resistance to what Bell is doing comes from a feeling that we must not move viability because it will restrict reproductive care. On the other hand, it is possible that some who support him do so in part because they *want* to see viability moved for the sake of restricting reproductive care. Bell and his colleagues would rather we consider their practice and their results on their own merits.

An alternative would be a system like Canada's, which does not have any legal restrictions on abortion; instead, these decisions are private and not intruded upon by legislation, as is the case for other medical decisions.

When I was in Iowa City to talk to Bell and his colleagues, I went along on morning rounds in bay one of the Iowa NICU for a few days. It's where the sickest fourteen babies in the hospital are sequestered. The youngest baby in bay one had been born at 24 weeks.

I hung back and watched the rounds, listening to the clinicians update each other on the babies one by one. The rooms, all private, were darkened, the light coming from the glow of the monitors. Each set of sliding glass doors was decorated with the name of the baby who lived inside. It struck me again how the energy of the NICU is so particular: the hushed darkness, the babies obscured under the technology. The well-meant decorative touches that gesture toward nursery-ness but seem only to make the un-nursery-ness more striking. The tightly controlled calm of it all, punctuated by swift, sharp emergency. The feeling of being in the in-between.

Every time I started to experience my reporting as an intellectual exercise, I looked at the parents, mostly mothers, sitting by the incubators.

They took in the jargon; they searched the faces of the doctors for a clue. Many had an air of restrained desperation. I wanted to wrap my arms around them. I saw, or imagined, how they felt stripped of the things that held them secure in the world—their jobs, their daily life—dropped, still bleeding, on this terrible, incredible postpartum Mars.

Bell's critics, like Dr. Joseph Kaempf, a neonatologist at Providence St. Vincent Medical Center in Portland, Oregon, feel that Bell may not be providing true shared decision-making at low gestations. This is because, at most hospitals that offer the choice of active care at 22 and 23 weeks, some substantial portion of parents opt for comfort care. At Iowa, the vast majority of parents of 22-week babies and virtually all parents of 23-week babies opt for active care. Kaempf feels that this fact suggests a subtle culture of coercion. In his own practice, Kaempf falls on the other side of the spectrum: He generally does not offer active treatment to 22-week babies (although he says it's not a nonnegotiable rule) and offers a choice of active treatment or comfort care at 23 weeks through 25 weeks. Many hospitals don't offer comfort care at 25 weeks in the absence of other problems beyond prematurity.

Bell doesn't have much patience with Kaempf's argument. In his experience, no one has ever said they wished he hadn't resuscitated their baby, although he also acknowledged that if a parent does feel that way, they may not tell him. "People say we are playing God," said Dr. Bell. "But you know that's what doctors do all the time, right?"

A few months after I returned home from Iowa, I spoke with Dr. Mark Mercurio, a widely published neonatologist and a professor of pediatrics and bioethics at Yale University. We were talking about how to counsel women in labor at 22 weeks. I asked him if he felt that resuscitation should be *recommended*, as is the practice in Iowa, rather than just offered. After all, I said, it seems like recommending resuscitation at that gestation—that sense of potential optimism—is one of the factors leading to nearly 60 percent survival in Iowa.

"There is a *chance* of survival at 22 weeks," he said. "So, does that mean that we should *recommend* resuscitation? I would say no. What it does mean, to me, is that we should make resuscitation available. Recommending is a whole different thing, because this is a long, hard road. Most of these kids are not going to survive. I'm amazed and astounded to hear that someone is reporting 50 percent survival at 22 weeks. I think you may be thinking of 23 weeks."

I opened the slides with the results from Iowa and read them again, out loud: 58 percent survival for all live-born 22-week babies.

"Amazing," said Mercurio, after a pause. "Those are extraordinary outcomes. Again, the numbers are low," meaning that it's hard to draw big conclusions when we're talking about only a small number of babies. "But even to say that somewhere between half and two-thirds, even if we're vague about it—that's certainly much higher than the experience of anyone else that I'm aware of. Doesn't mean it isn't true, but those are amazing numbers."

"Well, that's what I'm trying to make sense of," I said.

"I will tell you this," he said. "Dr. Bell, from my knowledge, has long enjoyed an excellent reputation in the world of neonatology, so if he said that these are his outcome data from the past ten years, I'm inclined to believe it."

Bell's data was unpublished at the time, so that likely explains Mercurio's unfamiliarity with Iowa's outcomes. Still, I was flummoxed by exchanges like this, because they suggested a deep schism at the top of the field in what is understood to be possible. What I didn't understand at first was that notions of the possible are, in fact, shifting all the time. Tucker Edmonds's framework of the possible versus the probable is a potent way to think about extremely premature birth—it is especially apt because it doesn't propose a single correct path or practice. That pendulum is dynamic; it requires thoughtful attention to each child, to each family, as well as to what the research says about the big picture. It asks that we live with contradiction, hold that contradiction, and move forward as best we can, even when there is no right answer. It's harder this way, but also richer. It allows for the possibility—though perhaps not the probability—of Alexis Hutchinson. A girl, flying.

13

Knowing When to Stop

DR. EDMUND LAGAMMA was sitting shoulder to shoulder with a young father; in front of them was a baby in an incubator, and arrayed around the incubator were an oscillator ventilator and several IV pumps and monitors. It was enough medical technology to fill a sedan. The baby herself was obscured under the tubes and wires; her small arms waved in the air, movements that at first glance looked typical. But her arms were jerky and stiff, and the longer I watched, the more obvious it was that something was wrong. "See," said LaGamma. "I think she's seizing."

A nurse practitioner hurried out of the room, silent tears streaming down her face. The father put his head in his hands and then he turned to LaGamma. "I know it's bad," he said. "I understand. But, please, can you just put a needle in and take the blood out of her brain?" And LaGamma began to explain again, for the fourth time in a half hour, that the bleeding in this baby's brain was the result of an asphyxia, lack of oxygen, at birth. She had had multiple strokes at birth and her brain was damaged; no one had a cure for this. He offered the father two choices: keep her on life support or let her go.

The air in the room was thick. I was standing in a corner, trying to be invisible. My agreement with the hospital—that I could observe and report in the NICU as long as I didn't disclose patient names or identifying information—suddenly seemed thin and vaguely shameful against the enormity of what I was witnessing. There were other babies in the room, which had space for six incubators. The room was one of several that make up the NICU at Maria Fareri Children's Hospital in Valhalla, Westchester County, New York. A few nurses sat quietly, watching other monitors, and one mother was doing kangaroo care behind a curtain, holding her baby to her bare chest. And there was this father, sitting in an ordinary hospital chair, trying to understand what LaGamma was saying and coming up against the brick wall of his own sorrow. They circled around and around: "Please take the blood out of her brain." "I can't take the blood out of her brain." His daughter was alive. If he allowed the doctors to withdraw life support, his daughter would die.

LaGamma, chief neonatologist at the hospital, felt that the kindest thing to do for this baby was to take her off the ventilator. He believes in trial of therapy and will recommend NICU care for 23-week babies and others who have long odds, like this baby girl, who was only a few weeks early but suffered a catastrophic complication at birth. The trial-of-therapy approach, by definition, necessitates a willingness to consider withdrawing life support later, if that treatment is likely only prolonging the dying.

LaGamma was careful to present both choices, but he also made it obvious what he thought: There was no effective treatment for this baby. He believed that it was acceptable and ethical to let this child go. He wanted the father to know that this can be a loving choice. A pediatric neurosurgeon quietly entered the room and sat down with them to help explain the damage to the girl's brain, the potential repercussions for her ability to ever breathe on her own, see, hear, move, speak, and understand. They circled back around. "Please take the blood out of her brain." "I can't take the blood out of her brain."

LaGamma has been having these conversations for the last thirty years, and he can recall the first time he had to withdraw life support from a baby—one Christmas Eve, decades ago—like it was yesterday. It hasn't gotten easier. Earlier that morning he had been talking to me about this girl's condition: Ostensibly he was explaining it to me, but also it seemed it was weighing on him. "A heart is beating and that heart is probably going to stop by the end of the day," he said. "And who caused it?" He pressed both palms to his chest. "Continuing treatment is intolerable for the child. It's not appropriate. This is the compassionate, sensitive, kind, and correct way to address the issue of futility. We've been running down the road. And it's a dead end. People want it to be 'natural,' but you wouldn't be here"—he gestured to the beeping, blinking, whirring battalion of technology surrounding us—"if this was natural."

This is true, and it's a great triumph of modern medicine that most preemies, upward of 90 percent overall, will go on to live long lives.

But not this particular girl. The conversation among the father, LaGamma, and the pediatric neurosurgeon stretched on. The father asked questions to which there are no good answers. LaGamma struggled to respond in a way that would provide clarity. Eventually, the feeling that I was invading the most private, nightmarish moments of this man's life became too overwhelming. I left and sat in the waiting room. Later I heard that the girl's mother had made it there to see her, that the parents had agreed to have their daughter taken off the ventilator, and that the little girl had died.

Stopping this baby's life support was probably the best thing for her. Still, I was sick for her short, sad little life, sick with knowing that I was one of the only people in the world who bore witness to her.

"It's excruciating," LaGamma said. "But you have to be commit-ted to your care model. Whenever possible, we give a trial of ther-apy, and then, if it is necessary, we withdraw care. That's much more labor intensive than having a hospital just say, 'Well, we don't resus-citate under 500 grams.' Or 'We don't resuscitate unless they're 23 weeks and 6 days.' Biology is messy: Young men don't all shave at fourteen. Young women don't all menstruate at thirteen. So, what makes us think that a baby at 23 and 6 is that precise in their bio-logical maturity?"

Withdrawal of life support is a uniquely modern phenomenon; it wasn't possible until intensive care was fully established for both adults and children in the last part of the twentieth century. Before that point, cardiovascular death was death. You stopped breathing; your heart stopped; you were dead. But once it was possible to keep a patient alive on a ventilator past any hope of getting better, the inevitable question arose: When should we stop and allow death to come?

As a medical practice for babies, withdrawal of life-sustaining care was first described in a controversial-at-the-time 1973 article in the *New England Journal of Medicine* by Dr. Raymond Duff and Dr. A. G. M. Campbell, both of Yale. In it, the doctors revealed that they had withdrawn or withheld active care for forty-three babies over the course of thirty months, which accounted for 14 percent of all deaths. It was the first time anyone officially admitted to such a thing.

In 1975 the Hastings Center, a bioethics think tank, published an interview with Duff about this contentious practice; the Q&A focused on questions that were new then and are familiar now, al-though we still to struggle to answer them. Like this one: Who de-cides that a baby can be taken off life support and by what criteria?

To the first of those questions, Duff proposed shared decision-making between parents and doctors as the best strategy. This is now the preferred approach, but it was a novel idea at the time, when medicine was still mostly paternalistic; doctors simply informed parents of what was going to happen, rather than engaging with them as partners.

Duff rejected the idea that giving decision-making power to parents would open the floodgates to poorly reasoned or unethical decisions. "It will only give the people who are most involved the freedom, the right, to influence their lives in ways that they think are most important."

The interviewer interjected, "If families are given decision-making powers, what is to prevent them from deciding quite arbitrarily that a child shouldn't be kept alive?"

Duff replied, "Physicians won't allow it. . . . Furthermore, families can be trusted far more than many realize."

But *which* babies should we consider removing from life support? Only the ones who are actively, certainly dying? Or is it acceptable to let babies die when they might live with profound neurological disability, with a quality of life that will be, in someone's opinion, worse than death? To this more difficult question, Duff gave an impressionistic answer—basically, that life support is justified when a baby has potential for "meaningful humanhood." Asked to define that term, he said: "Families usually consider that the capacity to love and be loved, to be independent, and to understand, anticipate, and plan for the future are important. This makes sense to most people." (It makes more sense to me to simply say "to love and be loved" and end it there.)

From the moment these nascent, dreamed-of technologies—in particular, mechanical ventilation—became modern intensive care, these dilemmas have been unavoidable. The debates haven't changed much, and answers are still elusive, although Duff's ideas about shared decision-making and the permissibility of withdrawal of life-sustaining care have clearly become the norm: Withdrawal of active treatment is now the most common cause of death in NICUs, accounting for about 60 percent of deaths.

The patient can remain alive (or "alive") when in fact treatment is failing or has failed or is overwhelmingly likely to fail. These dilemmas exist in adult ICU care, too, of course, but when the patient is a baby, the situation is even more difficult, because the decisions are made by surrogates, without any meaningful information about what the patient would want. The patient is unknowable. The patient doesn't have a living will. There are no remembered conversations to draw upon in making the decisions. (*I remember she said she would want everything done. I remember he said he doesn't want to live hooked up to a machine.*) And there is the heavy weight of this brand-new, unlived life hanging in the balance.

Dr. Brian Carter is a neonatologist and bioethicist at Children's Mercy Kansas City who has led the development of pediatric palliative comfort care. He remembers the days when doctors were not comfortable withdrawing life-sustaining care no matter what, sometimes even after death had, for all intents and purposes, occurred. He remembers horror stories from as recently as a couple decades ago; for instance, a neonatology fellow who was on his knees weeping, begging an attending neonatologist to take a dying baby off an ECMO (extracorporeal membrane oxygenation, i.e., heart and lung bypass) machine. "We would do everything to the point of no return. So a patient would literally die on life support, which is messy. Would die while getting chest compressions. Colleagues would never stop," he said.

From his residency in the early 1980s, he remembered one baby in particular: "Seven hundred grams, 25 weeks, four chest tubes in, and those were the four most recent chest tubes," Carter told me. "He had had a total of nine chest tubes put in his body up until that period of time. And he had grade three hemorrhages on both sides [of his brain], and he was in respiratory failure and there wasn't anything we could do, ventilator-wise. We didn't have any more magic. When he got bradycardic [low heart rate], we were told to do chest compressions."

Switching to a (perhaps self-protective) second-person perspective, he recalled: "You do chest compressions on this little chest and you're thinking, 'Wow. This is bound to be not good.' And ultimately you say, 'Okay, well, we tried for ten or fifteen minutes, and you've done this, this, and this. You can stop now.'"

He went on: "And then I had the 'pleasure' of going to the autopsy review and seeing what damage was done, and hearing the pathologist say, "Well, this is from the RDS, and this is from the baby's prematurity, and here is the hemorrhage in the brain. And this—this is from the chest compressions at the time of resuscitation: the bruising on the liver and the lungs. And it was like . . ." He let out a breath. "'Wait a minute.'"

As a result of experiences like this one, Carter helped develop the field of neonatal palliative comfort care, the science and art of helping babies die peacefully. He feels that withdrawal of life-sustaining care and initiation of comfort care doesn't happen enough, even now.

"Rarely does it happen that people redirect care," he said, speaking specifically of withdrawing life support *before* the baby is actively dying. "They just keep pressing on . . . I'm no psychologist or social scientist, but I think it's just plain human nature: The more you attend to and care for another person day in and day out, the more attached you become and you retain an interest in that person. And so, while some would argue that it's in the child's best interest to now redirect care, competing interests rise to the surface. The interests of the health care team, the interests of the family who is so hopeful. They're desperate; they're not going to give up."

As Carter points out, these practices, even now, are deeply imperfect: full of questions, differences of opinion, and grief. But generally there's been a culture shift in NICUs in the last decade or two, a recognition that we are in charge of the technology, not the other way around. There are exceptions, of course: the parents who, understandably, can't bear to say goodbye; the doctors who can't admit defeat. But it's less common now for premature babies to die on life support or with chest compressions than it was in the past.

For some parents, taking a child off a ventilator feels more active than never having put them on a ventilator to begin with, though medical ethicists say that the two choices are both acceptable. It is a small mercy that parents who have agreed to the withdrawal of life support generally feel that they have done the right thing for their child when asked about it later, especially if they think that the clinicians were caring and honest, and if they have seen their child's deterioration for themselves.

Laura and Graham Shullman, unfortunately, know this well. The couple, who live in a suburb of Detroit, longed for a child. After four years of trying unsuccessfully to get pregnant, they did IVF, which resulted in triplets. But Laura went into early labor as a result of an undiagnosed yeast infection in two out of the three placentas. Their children, Lilly, Stella, and Charlie, were born in 2015 at 23 weeks and 6 days' gestation. It happened so fast that there was no time for steroids, so all three of the babies had severely immature lungs.

From the beginning, all three struggled with multiple complications, but especially Stella and Charlie.

Around 2:00 a.m. on the day after the triplets were born, the neonatologist came to see the couple in Laura's hospital room. He said that if they were religious, they should baptize the babies right away. They weren't, particularly, but a NICU is like a foxhole. They called the hospital chaplain, who performed the baptisms.

"It was just a beautiful moment," said Laura in her endearing, flat upper Midwest accent.

"Over time, these memories change, right?" interjected Graham. The couple often finished each other's sentences. "At the time it was beautiful, but it was also terrifying."

"Devastating," Laura agreed.

"You're being told to baptize; if you've ever watched TV, that usually means shit's going down," Graham said.

"I remember sitting back in my wheelchair, bawling my eyes out," said Laura. "And Graham's crying. And Graham said, 'Please do what you can. We need to leave this hospital with a child. At least one.' And obviously you want all three of them. But at least one."

Charlie struggled. He had tubes inserted in his chest to relieve the pressure when his lungs collapsed. Then he developed necrotizing enterocolitis (NEC), in which part of the intestines dies. He had one surgery to remove a section of his intestine and recovered, and then his NEC got worse again and he needed a second surgery. After that, multiple organs failed. He was bloated from kidney failure, on the highest settings on the oscillating ventilator, but not maintaining good oxygen saturation. He couldn't recover.

"One of the heartbreaking things was that at that point we hadn't held him yet," Graham remembered. "So basically we made the decision to just take him off all of that stuff and let Mom and Dad hold him and just let him . . ."

"Not suffer anymore," said Laura. "I remember the doctor looking me in the eyes and saying, 'Laura, I know you comprehend what is going on. And I know you know what's best without me telling you.' Which was very powerful. He didn't have to say, 'It's time. End it.' That was his nice, sensitive way of saying, 'This child is not . . . If we prolong this, and he survives, which is, like, a 1 percent chance, he wasn't going to have any sort of life at all, really.'"

"At that point, we're feeling like seasoned veterans, but we'd only been there two weeks," Graham said. "We were always about hope, and anything can happen, and all that. But you know, it was not a question of . . ." He trailed off. "And it was a nice thing to be able to free him from those machines before he passed away, rather than to have him pass away still trying to pump air into his lungs. I thought the way we, in hindsight, we did the perfect—"

"The right thing," said Laura.

"The right thing to do," agreed Graham. "We got to say our goodbyes."

"I let Daddy hold him first," Laura remembered. "With the oscillator still on him. And then, when we took it off, I held him when he passed, which was the most beautiful experience I've ever had. I'll never forget it. I mean, it was beautiful and sad. I could feel his chest stop breathing on me, and it was something a lot of people never have to experience. You never think you're gonna go through this. But it was my moment with my son, the first time I ever got to hold him. And we're saying goodbye. And I'm just so thankful that it was me, who carried him, who got to hold him during his last breath."

"We had a day when we had to go to the funeral home and have

Charlie cremated and put him to rest," said Graham. "So you [can] imagine our day. You're in a haze."

"And you're just hoping you're not gonna come back two more times," said Laura.

Stella and Lilly spent 138 and 140 long days in the NICU, respectively. Stella had meningitis and NEC and endured two bowel surgeries. There were many moments when Laura and Graham thought they would have to say another goodbye. But, finally, both girls came home. They both have minor cerebral palsy, which resulted in a bit of a delay in their walking and some stiffness. Lilly also has autism. It is an extremely difficult road they are walking, but the Shullmans are so proud of their girls, who are now three and a half years old. And Laura and Graham are adapting, getting the girls the therapies they need, and enjoying them for who they are: a hilarious, tight-knit duo.

Both of the Shullmans say it was always clear to them that their babies might die, even though it was never really explicitly stated, The doctors didn't push stats on them; no one ever sat them down and told them they might have to take one or all of their children off life support. Instead, doctors said things like "He is very, very sick" and "Babies don't really bounce back from this kind of thing."

Some families might want more explicit, pointed information, but the Shullmans felt that the doctors conveyed what they needed to without overwhelming them. Laura and Graham felt empowered and included as parents.

Alexis Kropp-Kwon and her husband Young Kwon had a similar experience. Their son, Harris, was born at Pennsylvania's Lehigh Valley Hospital–Cedar Crest in March 2016 at 28 weeks as a result of preeclampsia in its most dangerous form, HELLP (hemolysis, elevated liver enzymes, low platelet count) syndrome, in which

Kropp-Kwon's blood pressure rose perilously high and her liver started to fail. At first Harris did relatively well, but when he was a month old his bowel became perforated, necessitating surgery; doctors removed the damaged portion of intestine. But in recovery from surgery, he had a major stroke. Doctors said that his brain had been damaged and he would likely have some degree of cerebral palsy.

"You have to mourn all these dreams that you had of your son coming home and being perceived as a normal, healthy baby," Kropp-Kwon told me. "We had to mourn everything we thought and we had to go forward. That's what we did."

So they waited for Harris's bowel to heal, and meanwhile he had to stay on the ventilator, stay on intravenous nutrition. Slowly, slowly, he improved, and when he was three months old, doctors started talking about when he might go home. Then he developed an infection and went into septic shock.

"It was bad. They said, 'Here are some options,'" Kropp-Kwon said, crying as she remembered. "My parents and my husband were like, 'We're going to do antibiotics, we're going to do this and that, we're just going to keep going the course.'

"I excused myself and went back into the NICU, and I said to one of the neonatal nurse practitioners, 'If Harris makes it through this, how much worse is the brain damage gonna be? How much worse is everything going to be?' She looked at me. The nurses were our family at this point. She said to me, 'It's going to be bad.' I said, 'Bad in the sense that I'm going to trick myself that Harris knows who I am, but Harris has no idea who I am?' And she said 'Yeah, that's a very strong possibility.' I walked out of the NICU and went back out to the waiting room, and I looked at my parents and I looked at my husband and I said, 'I'm not making any more lifesaving decisions for Harris. I don't want to bring home a baby. I want to bring home *Harris*."

"We all made the decision. We said if Harris starts to crash, we don't want any lifesaving measures. I believe this with all my heart: Science can prolong death. It wasn't fair to Harris.

"At 4:30 a.m. he crashed and we knew. We knew. That was our agreement: that if Harris showed us that it was time, then we were going to let him tell us. They put him in my arms and my husband was right next to me and we took him off his ventilator," Kropp-Kwon said. "At 5:00 a.m. I looked at the doctor and I said, 'He's gone.' He had passed. Then the most incredible thing happened. So tragic and so beautiful at the same time. I held my son without any wires or ventilation or any of that stuff. I held him over my shoulder. My husband held him. We cradled him. Every single nurse, every respiratory therapist, came and said goodbye to Harris. They held him like a baby, like a real baby.

"I looked at the head nurse and I said, 'What do we do?' She said, 'We're going to give Harris a bath and we're going to put him in his pajamas, and then you're going to say goodbye.'"

There is no silver lining to these losses; there is nothing that could be worth what these families have been through. But both the Shullmans and the Kwons felt supported and valued through the entire experience, as unbearable as it was. They both felt that the clinicians had stayed with them, hadn't abandoned them, even when it was clear that their babies were dying. Both families have found meaning and solace in staying connected in some way to the hospital where their babies had died and to the staff who had cared for them.

It doesn't always happen that way. Aleshia Jones gave birth to twin boys at 34 weeks in 2003 at Memorial Hospital in Gulfport,

Mississippi. The boys, Tyree and Jyree, were good-size, about 3 and 4 pounds respectively, but Jyree had respiratory distress and needed oxygen. A few days later he developed NEC and was flown to Ochsner Medical Center in New Orleans for surgery, which went well. But his health declined in the following days: He needed increasing respiratory support. Finally the doctors told Jones there was nothing else they could do.

"They allowed me to hold him until he died. He died in my arms. I held him as long as I could in that moment," she said. "Then they put me in another room with the chaplain; they brought Jyree in. I just looked at him and I would have sat there and held him until I couldn't hold him anymore. But I knew I had to let him go. I don't remember how long I held him. The chaplain prayed with us and finally I told them I was ready and I gave him back."

Jones felt that many of the nurses she encountered were empathetic but that the doctors were hardly present. She felt no one made a meaningful connection with her during her son's death; no one followed up with her about her mental health. In the months after Jyree's death, Jones struggled with postpartum depression and suicidal thoughts. Being present for Tyree, who came home after about a month in the NICU, was what kept her alive. Tyree—who is now a healthy high schooler who does Junior ROTC—used to dream of Jyree, missing the twin he never got to know.

In fact, Jones's experience is probably more common than the Shullmans' or the Kwons'. It is not unusual for clinicians to feel that the dying process is a time for them to step away, when there is nothing more that they can do. They might be uncomfortable. Or they might feel that end-of-life isn't really part of their job.

Dr. Elvira Parravicini disagrees. Her mission as a neonatologist is for every parent and every baby to be supported both medically and emotionally through the dying process. She says that doctors must be fully present during withdrawal of life support and not think of it as the moment when nothing more can be done. End-of-life doesn't mean that medical care is over; it means the goals of the care have changed, from life prolonging to comfort—comfort in every sense, for the baby and the parents. This is how she came to launch and direct the extraordinary Neonatal Comfort Care Program at Columbia University Medical Center/New York–Presbyterian Morgan Stanley Children's Hospital. Essentially, it's hospice for babies.

Parravicini was born in Milan, Italy, in 1956 and knew from a young age that she wanted to be a doctor for babies. "I *loved* them," she told me in her mellifluous accent, sitting behind the desk in her small, neat office, which is filled with photos of other people's babies and children. She is slender, with close-cropped hair and a calm, self-contained demeanor, like an ascetic. "At the beginning of life, there is so much hope. I really feel the injustice when it is a baby who is sick. And so I was going to save all of them." She smiled wryly and turned her palms up in a half shrug.

When she moved to the United States to work at Columbia, the NICU was twice the size of what she was used to back in Italy, and the population included many more babies who were on the brink between life and death—babies with serious cardiac conditions, chromosomal syndromes, and birth defects, and those who were extremely premature—possibly too premature to live. These babies raised new and painful questions for her.

"When I was very young, if you would have told me I would specialize in comfort care, I would have said, 'No way.' I would have wanted to intubate even at 20 weeks' gestational age," she said. But her youthful certainty was giving way to a different understanding. She wrestled with what to do for these most liminal babies.

"When I was in Italy, I barely saw this kind of baby. So I had no idea what their life would be like. Even just in the hospital," she told me. "Some of them were really, really, really premature. They go through so much, they just suffer, suffer, suffer. I mean, day in, day out. And it's true we have medication, we have painkillers, but you cannot just sedate them twenty-four hours a day."

Parravicini wondered if less invasive treatment—care that might not prolong life but would ease and support it—should be offered much more frequently. It is often too difficult for parents to bring this up themselves, so doctors must be the ones who make it clear that comfort care is an acceptable option.

These questions were on her mind when one day in 2008 she started her program almost by accident: At a staff meeting there was a discussion of two babies with Trisomy 18, a chromosomal abnormality that causes a constellation of critical medical problems and severe developmental disabilities. Only about 10 percent of these babies survive their first year, and when the diagnosis is made prenatally, the pregnancy is often ended. But two mothers had declined that option, and now the question was: Who was going to care for these babies, who were unlikely to live long no matter what? And what was the most humane way to treat them? Parravicini volunteered. She felt that she could support their lives—keep them comfortable medically, feed them, empower their parents to hold and comfort them physically—without deploying all the heroic measures often involved in intensive care.

She didn't set out to establish a formal program, exactly, but over time her colleagues would call her when a baby was born with a life-limiting condition or extremely prematurely, and she realized somewhere along the way that she was now a specialist in neonatal comfort care.

She has built certain protocols and tools: Parents can stay with babies at all times, even if they need medication. (In some hospitals, the administering of medications means that a baby must be admitted to the NICU, even if they are going to die quickly.) There are tiny boat-like cradles to make it easier to hold an extremely immature baby. There are special hats to put over the heads of babies who are missing part of their skulls. Cots with refrigerated mattresses can keep a small body cold, so parents can choose to keep their babies with them for hours after death. These are practical aspects of care that allow parents to parent their newborns for as long as they can.

The first time I met Parravicini, I asked her about a case that LaGamma, the doctor in Westchester, and his colleagues had written about in a journal. In the article, titled "Delivery Room Hospice," he wrote about a twenty-nine-year-old woman who came in with ruptured membranes at exactly 23 weeks. After counseling, the woman decided she did not want resuscitation and active NICU care for her baby. And so the baby girl was born and given to her mother; almost immediately she began gasping and heaving for breath in respiratory distress. This went on for hours. It was so unbearable for the mother to watch her daughter struggle that she begged for her baby to be taken to the NICU and ventilated. The doctors advised against it—the baby would have been suffering from lack of oxygen all that time—but they acquiesced. Seven days later, the girl developed large brain bleeds, life support was withdrawn, and she died.

In this article, LaGamma and his coauthors argue for delivery-room pain management for dying babies in situations like this one. Dying people of all ages gasp for breath—it is sometimes called air hunger—and preterm babies in particular are likely to have respiratory distress. It is not clear that it is painful per se, but it may be distressing, certainly to witness and perhaps to experience. Opiates are commonly used to make dying adults more comfortable; the fact that the medications may also hasten death is called the "double effect," and it is an accepted part of adult hospice care. All of this is much more controversial for babies; even the term "delivery room hospice" is a rough one. LaGamma told me he worried the article would be controversial.

For Parravicini, it is obvious that not only should the baby have been given any medication necessary for her comfort, but also that physicians needed to stay with that mother and explain to her what was happening and what could be done.

"No baby should die with suffering," Parravicini said. "We are in the modern age. We have painkillers that are really strong and powerful. And so it doesn't [have to] exist, this scenario, in which the parents should beg to intubate the baby because the baby is suffering."

Often, these conversations about allowing babies to die get tangled up in questions about projected future disability; this is complicated because, first, the projections are not always accurate. It is sometimes very hard to say with precision which babies will have minor or moderate issues and which will have severe, painful disabilities. And then there is the problematic and deeply subjective task of deciding if some disabilities are worse than death. Parravicini doesn't really engage with this. She thinks disability is not the most relevant factor. To her it's about the baby's suffering and whether or not all that invasive medical care is merely prolonging the inevitable—in other words, how the baby's suffering balances with the very real possibility that he or she will die no matter what doctors do.

"People always warn parents, 'Your child is not going to be normal, your child is not going to be good in school.' The vast majority of parents don't care. That's not the issue. To me, the issue is to see this poor little boy suffering for months and years. Even after the hospital. That's what matters.

"These babies undergo a lot of violence, in a sense," she said. "It's like, 'I'm not ready to be born and be alive.' That's what I feel sometimes. These babies telling me, 'I'm not ready to go through life.'"

She is not the only one to suggest that intensive care could, in some circumstances, constitute a kind of unintentional violence. In a landmark lawsuit, a Texas family sued Hospital Corporation of America (HCA) for resuscitating their 23-week infant, Sidney, against their wishes, partly on the grounds that the NICU care constituted battery of their child, who survived with severe disabilities that required around-the-clock home care. During the initial trial in 1998, Sidney was seven years old and had severe cerebral palsy: She could not walk, talk, sit up on her own, go to the toilet, or feed herself. She was nearly blind, suffered from seizures and cognitive disability, and needed a shunt in her brain to drain fluid. The parents won the initial case and were awarded $60 million, but HCA appealed to the Texas Supreme Court, which found that the hospital couldn't be at fault because it was required to treat the baby under a federal regulation that requires hospitals to treat any patient with an emergency medical condition, overturning the lower court ruling. This left Mark and Karla Miller, Sidney's parents, to struggle to pay for their child's lifelong twenty-four-hour care, and with the anguished worry of what will happen to her when they are gone.

The precedent set in this case applies only in the state of Texas. And even in that state, many hospitals and physicians interpret emergency medical regulation as requiring that a doctor *attend* to any patient with an emergency medical condition. Attending to the baby and giving them appropriate care could mean comfort care or it could mean attempting to resuscitate them. It is still quite common to give the option of comfort care at 23 weeks.

Since there is a lack of an agreed-upon legal or ethical framework—just a network of case law that sets precedents in certain states—neonatologists generally rely on what is widely considered best practice (essentially, the safety of the crowd) and on good communication and rapport with families to avoid conflict. Within the gray areas, though, there is still wide variability in both practice and belief.

The case of Sidney Miller shows that the lack of a consistent legal or ethical framework *can* leave families vulnerable to the overreach of medical authority. On the other hand, the lack of a set legal framework can also allow for the best and most compassionate medicine: considering every baby and every family as individuals on a case-by-case basis.

Parravicini has a radical commitment to the inviolability of the family, rooted in a belief that parental love and responsibility are unique; therefore, when there is medical uncertainty, it is parents—with input and support from doctors—who should decide what is best for their child. She often says things like "Allow parents to be parents." She believes that it is the acts of parenting that help people to remake their self-image into one of a mother or father. And that *feeling* of being a mother or a father is something that can be carried even if a child has died.

Princy Abraham would say that Parravicini helped her be the best mother she could for her first son, Aaron. One hot July day, I met Abraham at a Starbucks in Howard Beach, Queens. We sat outside in the sun. The planes landing at nearby JFK International

Airport roared low overhead, and Abraham's second baby, Aidan, napped in a stroller next to us. Abraham told me about Aaron's life, speaking through the tears that often ran down her face.

In 2011, Abraham was working as a neonatal nurse in the NICU at Long Island Jewish Medical Center. She and her husband, both immigrants from India, were overjoyed when she got pregnant. But at 24 weeks, Abraham's water broke. She was admitted to same hospital where she worked. A week later she gave birth to her son Aaron, who weighed 600 grams (1 pound, 4 ounces) and had a system-wide infection and severe respiratory distress.

"His lungs were horrible from the beginning," Abraham remembered. "He was intubated. I never got to hold him. And days went on. Months went on." Aaron's brain scans came back clear, and he eventually recovered from his infection, but his lungs were heavily scarred. He was diagnosed with severe bronchopulmonary dysplasia (BPD). This is a diagnosis associated with prematurity: It is roughly defined as needing supplemental oxygen of some sort after 37 weeks, an abnormal continuation of respiratory distress. In its mildest form, that might mean a baby being on a nasal cannula for a couple of extra weeks; or, in its severe form, it might mean a child who will never be able to breathe without a ventilator.

In Aaron's case, it was unclear if he would ever be able to breathe without help. Finally he graduated from the ventilator to a CPAP (continuous positive airway pressure)—that's the less invasive form of ventilation that delivers pressurized puffs of air into the nose and mouth through a mask, rather than through intubation. Nonetheless, he struggled to breathe as he reached six months of age, still hospitalized and dependent on respiratory support to live. "He was literally struggling, like a fish out of water, the minute the CPAP mask came off," Abraham said.

In every other way, he was a typical baby, smiling at his parents and meeting other neurological milestones.

"Everything was fine other than the lungs," said Abraham. "So I was stuck and the doctors were stuck. And I kept thinking, *If the lungs are not better at some point, he will be leaving me*," she said, meaning Aaron would die. "And why are they not offering that [comfort care] to me? I kept thinking it, but I couldn't say it aloud, because they would think I didn't want my child."

This left Abraham and her husband in a terrible kind of netherworld. Aaron's BPD was not improving, and there was no end in sight to his hospitalization. Abraham felt like a ghost. She didn't know if she'd ever be able to bring him home. Children who need long-term or permanent ventilation can sometimes come home with the machinery, if home nursing is available where they live. Sometimes, in a matter of months or years, they might improve and be able to breathe on their own. But sometimes they must spend long periods of time, sometimes their whole lives, in long-term care facilities, depending on their level of medical need.

Then Abraham heard that an experimental drug for severe BPD was being trialed at Columbia's NICU. If she could transfer there, maybe her son could get into that study. Six months after Aaron was born, he transferred to Columbia. But shortly after, his lungs worsened, putting strain on his heart, and it became clear that he couldn't make it any longer just on CPAP; he needed to go back on the ventilator. This meant he wasn't a good candidate for the experimental drug, either. Instead, he would need permanent mechanical ventilation through a tracheotomy—a "trach" for short—a hole cut in the throat. This allows for long-term ventilation while also allowing a child to learn to talk and eat, because there's no tube in their mouth.

"He was almost Aidan's size," Abraham told me, gesturing to her chubby, healthy, sleeping baby in the stroller next to us. "On CPAP, Aaron was happy, kind of. And now I'm going to lose him again. Get that breathing tube in him again. All these IVs and drips and everything to keep him alive."

It was at this point that she met Parravicini. "So when Elvira was rounding, she told me, 'Princy, his lungs are very bad. We need to trach him.' Trach him and take him to a long-term facility. And that was something that I couldn't accept."

"I asked her, 'What if I didn't want to trach him? She said, 'Okay, we will make him comfortable. We will do comfort care for him.' And that was a new world for me."

So they kept Aaron on CPAP instead of traching him. They gave him whatever he wanted. Princy held him constantly. "I'm in the hospital holding him, 24-7. Doing whatever I can. His heart got tired of working, working, working, and that was it. So within two weeks—" She stopped and was silent. She meant that within two weeks he died.

The comfort care that Aaron received included active medical treatments: The CPAP kept him comfortable and from experiencing air hunger; he had every medication he needed. His parents received counseling and support, both in the hospital and after Aaron died.

Parravicini remembers Aaron very well. "He had gone through so much by the time he got here," she said. "Oh, this beautiful baby. It was really tough, deciding not to intubate him." She acknowledged that providing comfort care to a baby like Aaron is not something that all hospitals, all doctors, would offer (although ventilator-dependent BPD is one of the conditions that bioethicists consider qualifying for neonatal palliative care). After all, it would have been possible to keep him alive, at least for some uncertain number of weeks, months, or years. But Parravicini said Aaron's condition meant a certainty of suffering, a life that would be spent in and out of hospitals and institutions.

"In our experience, when they have BPD and you put in the trach, they develop pulmonary hypertension and they end up going from the NICU to the PICU [pediatric ICU]. Most will die. Some will go home or to a facility for a while, and then they come back to the hospital. But they're not going to have a long life." She thought for a minute. "We did have a few cases like this in which the parents were, like, 'Yes, we want to go ahead with the trach.' We are happy to serve those cases. It's really, at a certain point, it's a parental choice."

The questions are endless and in some ways unanswerable. But parents don't see a gray area or an ethical debate. They see their one, particular, irreplaceable child.

Laura Shullman spoke to the truth of this. She said that Charlie was telling her and Graham everything they needed to know. It wasn't a matter of agonizing over which action to take but of watching Charlie and recognizing when he couldn't keep going.

"I knew in my heart that he couldn't struggle anymore," Laura said. "He was so sick. He didn't have any more energy to give. I was grateful that it was our decision. You just know in your heart when it's time."

14

Choice, Decisions, and the Messiness of Real Life

ANNIE JANVIER WAS in labor at 23 weeks and 3 days. She lay in a bed at the Royal Victoria Hospital in Montreal, now part of the McGill University Health Centre; her cervix was dilated four centimeters and her amniotic sac was bulging, threatening to break. Maybe she could stay pregnant a little longer—a few days, a week? Delay was the best she could hope for. As it was, she was in a gray zone, a period of time in which this hospital held that she could choose to have doctors attempt to resuscitate her daughter at birth or she could opt for palliative care. When she tells this story now, she pauses to say how grateful she is that she was in a hospital that offered her those choices. No one knew exactly when the baby would be born, but it was very likely going to be catastrophically early, and Janvier needed to think about what she wanted doctors to do in the delivery room. Someone handed her an information sheet that listed statistics for babies born at extremely early gestational ages and at extremely low weights: the percentage that survive, the percentage that die, the percentage that are blind or have cerebral palsy. The numbers swam in front of her eyes, and she had a welling up of rage and dark laughter. She felt as if she were in a Pedro Almodóvar movie.

She hardly needed to read the sheet, because she had actually written it herself. She was a neonatologist at this very hospital and was in the midst of completing a bioethics PhD on decision-making in the NICU. Her husband, Dr. Keith Barrington, was also a neonatologist. But it wasn't until she had that paper in her hand, looking at it with new eyes, that she fully realized the numbers were not effective signposts on the way to a decision, and that they were not helping her navigate this terrain. The numbers were a language from another life—a life in which she thought that the more scientific information one could provide to families, the better.

From where she sat now, in a hospital gown instead of scrubs, the information sheet was cold and infuriating; it was pessimistic, incomplete, and inadequate. Janvier knew how to thread an endotracheal tube down the tiniest of airways. and she knew how to calibrate glucose levels and watch for the first signs of sepsis. She knew all the nightmarish things that could go wrong for her daughter and exactly how unlikely it was that everything would go right. But now that she was a member of this unlucky club, a new resident of the weird and terrible netherworld where the imagined future evaporates, she understood that statistics alone were not going to help her do the right thing for her child.

"Sure, I had all the numbers. But it was what numbers *meant* that was important," she told me in her French-accented English, thirteen years later. For example, she asked me to imagine that I needed to have my arm amputated: What would I really want to know? "What *is it* to have an amputation? Can you find a boyfriend? Can you change hands to write? Are you able to drive? Do you make friends? How do amputees work in the world? Are they happy again? That's what a person who's going to lose their arm thinks. Not necessarily 30 percent this, and the prosthesis is like that," she said. "It's kind of the same thing as: How will I be happy if my baby's disabled? Will I feel guilty if I stop the respirator, thinking maybe she could have survived? If she's disabled, can she still have a good quality of life? Will she at some point look at me and be mad at me?"

Janvier acknowledges that the numbers are useful to a point, for context. But most parents are just hoping their baby lives. "Your baby's 100 percent dead or 100 percent alive" is how she puts it.

"Knowing all the percentages, you can't decide. What helps you decide is your approach to regret, how you live with yourself, your life experience, the dynamic within the couple. It's things that nobody speaks about. They speak about numbers and wanting to have decision aids on how to convey to parents what 30 percent means. But people don't decide rationally. They decide with not what they might gain but what they're afraid to lose."

I am a member of several preemie parent support groups on Facebook. Many of the posts are from parents with babies in the NICU who have just gotten some bad news. Inevitably, what they ask for are stories of others who have been through the same thing: What happened to your child? What's your life like? They're looking for hope, of course, but not just hope: They're looking for something solid to hold on to, a future to imagine.

Recently, the mother of a 23-week baby who had just been diagnosed with a grade 4 brain bleed posted. "I need to hear stories, please," she wrote. The responses she got were the texture of real life: Here they were, other kids who had suffered similar catastrophes. Some of the kids had serious cerebral palsy, some had mild CP, some had minor delays, and a few had no issues whatsoever. One in a wheelchair, one with a cane, a few with leg braces, many of them with glasses, many of them with shunts in their brains to drain fluid. They were playing in the snow and swimming in a pool. They were making silly faces; they were smiling hugely; they were snuggling with their parents. Their parents wrote of physical, occupational, and speech therapies, surgeries and struggle. But they also said: *She's doing everything the doctors said she wouldn't. They said she would be a vegetable. Don't give up. Don't underestimate them. This is my sweet, brave girl; this is my sweet, brave boy.*

It's not that anyone can give a definitive answer, but they can share the spectrum of life that emerges on the other side of all that uncertainty. It's the reply to Janvier's question: What *is* it to have a 23-week baby with a grade 4 brain bleed? Well, it's all this, and more. The only possible outcome missing from the thread is death. These support groups don't shy away from death, although the dominant narrative tends to be one of survival against the odds.

Research suggests that parents often want guidance and advice from doctors in these scenarios. They want specifics, they want stories, they want to know what the doctor would do if it were *their* baby. But doctors tend to give "objective," impersonal information instead of engaging in that kind of partnership.

Shared decision-making, the current standard model in health care, sounds egalitarian and logical. The idea is that the model respects both the diversity of values held by families and the special knowledge, experience, and ethical framework that doctors can offer. Unsurprisingly, this is not as simple as it sounds. It's simpler just to tell a family what the decision is—the old model of medical paternalism—or to swing in the opposite direction to full patient autonomy: lay the stats on the parents, give them those info sheets, and tell them they've got to decide on their own. It is much, much harder to do what Janvier is suggesting: To engage deeply with the family, to ask what is important to them, to talk about lived realities as well as outcome statistics.

Janvier managed to delay giving birth until her daughter, Violette, was 24 weeks and 5 days. At that point, with the benefit of steroids and the extra week of gestation, her decision tipped clearly toward active treatment. At that point, a girl who has had steroids has a greater than 50 percent chance of survival. She is now a beloved and active thirteen-year-old who has conquered some learning disabilities. Her life has informed her mother's practice of neonatology and medical ethics; in a way, it has made her into an activist on behalf of preemies.

Janvier does not pretend that all babies can live or that redirection to palliative care is never appropriate. She is not at all a treat-at-all-costs partisan. But she does feel that some clinicians have an unjust bias against preemies. For these babies, initial attempted resuscitation is often considered optional when, for patients in other groups with similar chances of survival, it would be considered a given.

Janvier posits that doctors are so worried that treatment will result in a child with a disability that they push the worst-case scenarios on vulnerable parents, hoping to discharge this sense of responsibility. They focus on the negative aspects of disability in ways that they would not do if the patient were already disabled. She is horrified by decision aid sheets made for parents that have illustrations of children sitting all by themselves in wheelchairs, implying that life with a disability is nothing but a terrible, isolating burden on both the child and their family.

"They say, 'Oh, your kid had a brain bleed. Blah, blah, CP. Blah, blah, blindness. Blah, blah, rehab. Blah, blah, special school.'

Nobody says, 'Blah, blah, *happiness.*' Nobody says anything about how your kid will love you and you'll love your kid. About what really counts in these situations: the resilience, the adaptation, having a family, having resources."

A study by Janvier; her husband, Dr. Keith Barrington; and Dr. Isabelle Leblanc provocatively titled "Nobody Likes Premies" is one of several papers that show how premature babies are treated differently than other patients. In a questionnaire, practicing physicians and university students were presented with the stories of eight patients, all of whom were currently unconscious and not breathing and who needed emergency resuscitation to survive.

The eight patients were all of different ages and had different illnesses or injuries. There was, for example, a two-month-old with meningitis; a seven-year-old with multiple disabilities, including deafness and cerebral palsy, who had just suffered a critical head injury; a fifty-year-old who had just been in a car accident; and an eighty-year-old with dementia who had just suffered a stroke. And there was a newly born 24-week infant.

The patients were given chances of survival that ranged from 50 percent to 5 percent, and they all had varying risks of being left with a disability if they survived. Notably, the 24-week infant was one of the patients with the very best odds of living and of living without a disability—a 50 percent chance of survival with treatment and a 50 percent chance of having a disability if they survived.

Study participants were asked if they would resuscitate each of these patients and consult intensive care for admission "always, generally, exceptionally, or never." They were also asked in what order they would resuscitate the patients if they all presented at once.

The results of the study are astonishing and are telegraphed by the title of the paper. Essentially, despite their superior odds, premature babies were not treated like other children. All the other children were given priority for treatment, regardless of existing disability or potential for resulting future disability. This is a common ethical framework: The deaths of children are particularly wrenching because they have so much life to lose. But the premature baby was given second-to-last priority for treatment, just in front of the eighty-year-old with dementia, despite being the youngest patient of all and one of those with the best odds.

"The life of the preterm baby was quite clearly seen as having less relative value than others in ways that are not explicable by established ethical frameworks," Janvier and her coauthors wrote.

In the absence of differences in outcome odds, the reasons for treating premature patients differently from others may be emotional—a deep, unconscious feeling that premature babies aren't really "here."

We are only about a century removed from a time when many parents would have experienced the death of at least one baby. It's possible that we are somehow evolutionarily programmed to expect the deaths of infants, an effect that would be magnified with a fragile or too-small infant. Maybe we developed a self-protective mechanism to designate babies like this as not fully human. We see them as having one foot in the mystery, in the void, almost as much as an eighty-year-old person who isn't oriented to reality. In other times and in less industrialized countries, these babies would be classified as miscarriages. To act to save them feels different from acting to save other patients. But should it?

In another paper Janvier and Dr. Mark Mercurio wrote that when a two-year-old is resuscitated and treated in the ICU, we want that child *back*, we want them back the way they were before they got sick. In the case of a premature baby, we don't want them back to how they were before; we want them to grow and be healthy. But we have no reference for that baby as a person, no web of relationships and connections.

In these cases, doctors may feel that they are not saving a life but creating one—and not only that, but creating a disability, rather than seeing the disability as a sometimes unavoidable side effect of treatment. Very premature babies' brains are unformed to such an extent that they lack the eloquent functions that we think of as making us human. So when those infants are tended in a NICU, in some sense they are being gestated in a NICU. In the process, the treatment can harm them—particularly harm their brains—while also saving their lives. But that is true of many medical treatments.

Janvier and Mercurio put it like this: "Some physicians even refer to a preterm who survives with disabilities as a 'product of the NICU.' It is interesting to note that one less often hears a disabled child referred to as a 'product of the Emergency Department . . .'"

Janvier gave me the following example to illustrate the point.

"If a child comes into the emergency room with a severe burn, his body is 70 percent burned, we don't start to show videos and do an informed consent and decision aids to the parents before intubating the child who inhaled smoke. We don't say, 'He's going to look like this. He may become addicted to morphine and it's going to take a long time to withdraw. He's going to look like crap: *This* is what he's going to look like. People are going to laugh at him. He may have sepsis. He'll stay in the hospital for a year. It really hurts.' Then go through the three pages of everything that may happen to that poor kid and say, 'Well, do you want us to intubate?' No. They don't do that. They intubate. They give the pain meds. And if the child develops sepsis, they say, 'Shit, it's not going well,'" she said, acknowledging again that not every story has a happy ending.

"I think it's important for parents to understand and be part of what's going to happen to their babies," she said. "And some parents don't want intervention, and you have to respect that, too. But quickly you can see which parents don't want to see all your information sheets and decision aids and hear about cerebral palsy. It's just too much information."

Janvier often makes the point that "quality of life" means different things to different people; in another essay she wrote of a 27-week baby who had a brain bleed. He was stable for the moment but critically ill and was expected to have some degree of cerebral palsy if he survived. The family had questions, but not the ones Janvier was expecting: They wanted to know if their child would be able to love them and be loved by them. They wanted to know if he would be capable of having sex someday. And they wanted to know if he could put pepperoni on pizza. It turned out that they owned a pizza shop, and they were wondering if he could work at the family business. To them, that's what constituted a good life.

What's important to parents might not end up being important to grown children, but until the children grow up, the parents' values must suffice. To some extent, notions about quality of life are culture-specific, but they are also idiosyncratic and intensely personal. Dr. Dagle, in Iowa, told me about a particular Amish family whose biggest concern was that their son be physically able to participate in farmwork. Then again, it would be a mistake to look at every Amish family and assume that physical ability is what matters to them above all else.

When providers assume that they know what certain families need or value, it can quickly lead to scary moral territory. Horrifyingly, one neonatologist told me that she heard from a fellow neonatologist that, at the tip-top-tier Northeast hospital where this person works, some doctors were reluctant to resuscitate borderline viable infants because the hospital serves many Black families, and the doctors felt Black parents would be less likely to agree to taking a dying baby off life support later if necessary. (There is some evidence that Black people are generally less likely than other racial and ethnic groups to consent to hospice care. In some cases, this may be because they, quite rationally, fear that those in authority may not value their lives.)

And sometimes I think we are focusing on the wrong questions. What does choice and shared decision-making mean in a society that doesn't guarantee access to health care, where coverage of pre-existing conditions—like prematurity itself, and all those conditions that arise from prematurity—is constantly in jeopardy? Where parents can't get time off to care for their kids with complex medical needs? Where special education services are constantly being cut?

Of all the children I have seen in a NICU, one of them stays with me more than any of the others. I'll call her Emma, although that isn't her name. She was at the University of Iowa NICU, and Dr. Klein wanted me to meet her, because he said she had the only case of severe bronchopulmonary dysplasia (BPD) in the hospital.

We walked into a small, sunny room. In the bed was a little girl who had just turned one year old; she was still in the NICU more than 365 days after her birth at 24 weeks and 425 grams, just under a pound, small for her gestation. She was wearing a teal onesie and was sitting propped up, on pink camouflage sheets. Two balloons were tied to her bedpost for Valentine's Day, one in the shape of a monkey and one a heart, printed with "I love you." A Cars poster was on one wall. She had a mobile set in front of her, with colorful tactile toys that she could reach. She was attached to a ventilator through a hole in her throat, a trach.

Emma was neurologically typical for her age: a one-year-old, in that sweet spot between baby and toddler. But her lungs were badly damaged. In this way, her condition was similar to Aaron, Princy Abraham's child.

Emma was shaking her head back and forth, back and forth. I tried to talk to her, the way you talk to a one-year-old—"Hi, sweetie, is that your bear?"—and she glanced at me but kept shaking her head, back and forth, back and forth. Klein said Emma's mother very rarely visits and her father never does.

Klein looked at Emma, who was still shaking her head. "She's, like, 'No, don't give up on me. No, I wanna live.' Emma, should we stop your ventilator?"

He said this for my benefit, to make a point, because we had been talking about how Iowa has a more interventionist approach than other hospitals; Klein feels that many other institutions are overly pessimistic about preemies.

Emma was too young to understand any of it. Still, what he said made me extremely uncomfortable on her behalf. I didn't know what to do, so I changed the subject. A few minutes later Klein motioned for me to come with him. "Bye, Emma," he said as we left the girl alone in her room, still shaking her head back and forth. I asked if she would be able to go home soon. "No," he said. "We've gotta get her on a home ventilator, which takes a couple of months."

Later, in his office, Klein told me how much it bothers him when people suggest Emma is in pain. "Other people say, 'Oh my God, she's suffering, she's on a ventilator. She's being fed with a tube in her stomach and she's got a tube in her airway, she's suffering.' No, she's watching TV."

"But do you think she's lonely?" I asked.

"I think she is," he said. "And we try to have volunteers coming in all the time."

Emma's innate human worth is not dependent on her ability to breathe without a ventilator, and there are kids who, with lots of support and therapies, do live on ventilators for years. Nevertheless, I found her solitary situation much sadder than Klein seemed to expect. That might be because I don't have all the context. Klein has been caring for her for a year, whereas I was with her for ten minutes. He feels that what she has been through will be worth it, because he expects she will be able to get off the ventilator someday.

This kind of outcome, a ventilator-dependent child with a tracheotomy, is not at all unique to Iowa; it happens everywhere premature babies are treated. And it's hardly the hospital's fault that they can't nurture Emma as a parent would. I don't know why Emma's parents don't come often, but I can think of all kinds of reasons: Maybe they live far away and don't have a reliable car. Maybe they have to work long hours to maintain their health insurance and pay the rent. Maybe they have health problems themselves. Maybe they have other kids. Maybe they can't bear it.

I asked Dr. Brian Carter, the palliative comfort care specialist, about the questions raised by ventilator-dependent kids with long-term tracheotomies. Carter explained it this way:

"Until the child can go home or until they die, they would live in the hospital. Some have lived in the hospital for three years. Is that good? Is that in the child's best interest? Well, probably not. It totally makes one reconsider what it means to be a child and why we care for and love children."

Then he told me about a bright spot, the kind of care he wishes every child who needs it could have. The Children's Center Rehabilitation Hospital in Bethany, Oklahoma, is an inpatient center that provides residential care for kids who can't live at home because of complex medical needs like long-term ventilation. Carter said that it struck him as unique in the way it lets kids be kids. They have aquatic therapy, pet therapy, music therapy, various classes. In some cases parents can sleep over. It is bright and cheerful. How many families have access to this kind of facility? Not many.

It's important to deeply consider the use or withholding of life support for extremely premature babies; it's important to examine our beliefs about quality of life and death, and to acknowledge how subjective and personal they are. It's important to ask clinicians to be partners who can engage on a human level, not just be providers of data.

But if we decide, as a society, that we should use the technology, that we want to say yes to all that hope, we should also decide, as a society, to support all the possible outcomes of that choice. This would ask much more of us. As it is, if a child is left with complex medical needs, it is up to the family to figure it all out. The level of hardship involved can be staggering. Sometimes it means that a toddler on a ventilator lives in a hospital bed all by herself. What do we owe babies like Emma? I think we owe them everything.

Part VI

The Crisis

The Body Under Stress

15

Racism Causes Preterm Birth

IF YOU NEED a reason to feel hopeful about America, I can recommend traveling around the country to talk to the people who take care of the tiniest, most vulnerable Americans. But if prematurity illuminates the extraordinary, it also reveals cruelty embedded in our society. When you consider which communities suffer the most, it becomes clear that unjust inequalities drive preterm birth in a direct and literal way. We can't understand premature birth if we don't understand why some people experience it more than others.

The United States as a whole is suffering such a high rate of prematurity that it qualifies as a public health crisis. Despite our advanced technologies and resources, the United States is the most dangerous place in the industrialized world to have a baby, both for mothers and their children. On the face of it, it's hard to understand.

Researchers and public health experts who work with the communities hardest hit by prematurity have insight into this, and it adds up to one big picture: American women are stressed. And study after study has shown that clinically significant stress is a major risk factor for preterm birth. "Stress" does not mean that pregnant women should just calm down because getting upset is bad for the baby. This is not about individual women who need to do some deep breathing. It is about societal forces that leave women and families hung out to dry—and some much more than others.

In the clinical sense, "stress" does not mean a tight deadline, a demanding boss, a traffic jam, the PTA meeting getting moved. (Normal life can be demanding, but it's just not a risk factor for preterm birth.) The kind of stress that matters here means grappling with serious mental and physical illnesses in yourself or close family; income, food, and housing insecurity; complicated grief; unsafe working conditions; the threat of gun or police violence; lack of access to health care and child care. These are issues that hit some American women hard. For these women, there is a lack of a social safety net; there is a lack of a sense of safety.

This leads to predictable and measurable hormonal and cardiovascular effects on the body. Scientists call this "allostatic load"—a measure of the hormones that the body releases in response to stress. These hormones, like corticosteroids and adrenaline, are meant to help humans deal with life-threatening situations. In the short term, they protect us by allowing us to respond quickly to danger, acting on our bodies and brains in myriad ways: raising our blood pressure, altering our metabolism, and sharpening our attention. But when these hormones flood our system too often over the long term—like a revving engine flooded with gasoline—it is harmful.

The exact mechanisms behind this are still being studied, but high allostatic load is associated with cardiovascular disease, hypertension, and other health problems that in turn can lead to a premature birth. Or a stress-associated hormonal and immune response may bring on preterm labor or other problems that necessitate preterm birth. Preterm birth can be a physiological consequence of living with too much anxiety and fear. A body under this kind of stress diverts its resources to survival; it isn't sure it can afford a pregnancy.

Among these stresses, there is one very specific kind of stress that is a main driver of the high preterm birth rate in the United States: the stress that comes from living with racism in its many forms, particularly racism against Black women. In fact, studies have shown over and over a link between the experience of racism and giving birth early.

On medical lists of risk factors for preterm birth, Black or African American race is often cited. But that is not accurate: It's not an inherent risk somehow conferred by skin color; it's a risk that comes from environment and experiences. But pinning the blame on the color of a woman's skin has sunk into our collective consciousness. Even, in some cases, the consciousnesses of Black women themselves.

When Dr. Joia Crear-Perry gave birth to her son at 22 weeks in 1996, she blamed herself. Many mothers feel this way—*Was it because I picked up that heavy box? Was it because I drank coffee? Was it because I was on my feet all day? Because I worried too much? Because we had sex that night? Because I gained too much weight?* But Crear-Perry was worried about something more pernicious—a feeling that lay blame not on anything she'd *done* but on who she *was*. She knew that as a completely healthy twenty-four-year-old woman with a planned pregnancy, she didn't have any risk factors—none at all—except one. "I believed it was my Blackness," she told me, more than two decades later. "I assumed there was something inherently wrong with me because that's what I was taught."

Black American women give birth early about 50 percent more often than other American women. The average preterm birth rate in the United States is 10 percent and it breaks out by race and ethnic-ity roughly like this: Asian and Pacific Island women, 8.6 percent; white women, 9 percent; Hispanic/Latina women, 9.7 percent; Na-tive American women, 10.8 percent; and Black women, 14 percent. (These categories are blunt: people of Asian and Pacific Island de-scent, for instance, are hardly a monolith, but this is what we have for national data.) The disparity is even more dire when you focus on very premature babies or those born before 32 weeks, who are at highest risk for death and serious health problems. Black women are almost three times more likely than white women to give birth before 32 weeks.

The stark consequence of this is that Black children are more than twice as likely to die before they turn one than white children, mainly because of prematurity. Here and throughout this chapter, I am comparing Black people to white people *not* because white people represent some sort of default or baseline but because these comparisons can help illuminate a problem.

When I mention this disparity to people who are just learning about it, the first thing they often guess is that it might be because Black women are more likely to be low income. It is true that income inequality disproportionately harms Black communities—and that fact flows from racial discrimination, too. And it is true that being low income is *also* an independent risk factor for preterm birth for women of all races and ethnicities. But for Black women, even after you control for socioeconomic status and education level, the higher rate persists.

A healthy, nonsmoking, affluent Black attorney in her early thir-ties living in a fancy apartment in New York City within a stone's throw of great hospitals is at least 50 percent more likely to give birth early than her identically healthy, identically affluent white neighbor down the hall. Put another way, Black women who get prenatal care starting in their first trimester are at higher risk of preterm birth and other complications than white women who do not get any prenatal care at all. Black women with master's degrees are at higher risk than white women who did not graduate from high school. In other words, for non-Black women, gaining the advantages that one would expect to be good for birth outcomes are, indeed, good for birth outcomes. Gaining those same advantages doesn't help Black women nearly as much.

So this disparity is actually not driven by income, education, or access to health care. Research has shown that it is also not driven by behavioral factors like alcohol or tobacco use. So, could there be something fundamentally different about Black women genetically that makes it more likely they will give birth early?

For years, despite mounting evidence to the contrary, this idea has been embraced by some researchers and clinicians. There is a sizable investment in the idea that race represents somehow an inherent and independent risk factor for preterm birth in and of itself. Hence the existential self-blame that Crear-Perry felt when her son was born: *I thought it was my Blackness.* That's what she had been taught in medical school. But this idea is deeply flawed: People who are perceived as Black do not necessarily have similar genetic makeups or innate health profiles. What we call race has social, political, and cultural importance, but it is not a meaningful genetic category.

Genetic and biological differences do not naturally fall along skin color lines. The genome is 99 percent the same in all humans. And the small amount of variation within it does not neatly correspond to "racial" groups as defined by skin color and other physical features. In other words, any two Black women will be as different from each other genetically as they are from any one white woman, and vice versa.

This means that skin color cannot be used as a proxy for understanding someone's genetic vulnerabilities or general health profile. There is no such thing as a disease that neatly corresponds to skin color. What about something like sickle cell anemia? It turns out that people who get sickle cell anemia have two copies of a gene that evolved in areas where, historically, mosquitoes carried malaria. (The sickle cell anemia gene confers protection against malaria.) That means that people with ancestors from certain parts of Africa are more likely to have sickle cell anemia, but so are people from places like South Asia, the Middle East, Greece, Italy, and parts of Central and South America. It is not simply related to being what we call "Black" or "African American"; it is much broader and more complicated than that.

New research suggests there is genetic contribution to preterm birth—but the important distinction is that there is no evidence that it explains the racial disparity, and no reason to think that genetics will ever explain why Black American women suffer early birth more often than others.

There's one kind of research on this issue that is particularly illuminating, and devastating. These are studies that compare preterm birth and low birth weight rates between American-born Black women and foreign-born Black women who emigrate to America from countries in Africa or the Caribbean. This kind of investigation was pioneered by Chicago neonatologists Dr. James Collins and Dr. Richard David, starting in the 1990s. There have been several published over the years, looking at different populations.

These studies have all overwhelmingly demonstrated the same thing: White American women and Black immigrant women have very similar rates of preterm birth and low birth weight. Black women born in America—including the immigrants' own daughters—are the ones who experience the higher rate of low birth weight and prematurity.

Collins, who is a professor of pediatrics at Northwestern University, wrote in 1997: "Regardless of socioeconomic status, the infants of Black women born in Africa weighed more than the infants of comparable Black women born in the United States." This is despite the fact that American Black women tend to have more mixed ancestry than African Black women, so if genetics were really at play here—if African ancestry were the source of the problem—you would expect exactly the opposite of what this research shows.

These studies strongly suggest that the source of the risk is not anything innate; it is experience in the United States. This research says something unbearably sad about the promise and hope of this country versus the reality of it. Immigrants come to America in the hope that their children will have a better life—but the daughters of Black immigrants have, on average, worse birth outcomes than their own mothers. By contrast, on average, the daughters of white immigrants to America have bigger, healthier babies than their mothers.

"I think Dr. Collins's research has pretty conclusively proven that the issue is what those babies and their families experience throughout their lifetime *as a consequence of being in America*," said Dr. Arthur James, an obstetrician-gynecologist at Ohio State University. "It is the impact of that differential treatment for those of us of color that has a detrimental impact on our physiology and places us at increased risk for preterm birth and infant mortality."

The exact biological pathways by which exposure to discrimination manifests in early birth are complex and still not fully understood. We do know that a high allostatic load—that stress hormone reaction—is central to the way that chronic stresses, like exposure to racism, harms the body. In 1992, public health professor and researcher Dr. Arline Geronimus coined the term "weathering." Weathering is essentially the idea that cumulative stress on your body over time causes health problems. This happens to everyone with age. We all wear out. Generally, the bodily burden of pregnancy is not as easily carried the older we get. But Geronimus hypothesized that Black women experience negative pregnancy outcomes at much younger ages than non-Black women because their bodies have weathered faster as a result of being exposed to more stressors. This follows to a sadly logical conclusion: Black women also have shorter average life expectancy than white women.

Perhaps some of the initial skepticism that this information meets has to do with the fact that, especially for people who don't experience it, racism seems conceptual and nebulous, difficult to measure. But there are ways to look at the effects of racism in a scientifically rigorous way. The effects of racism are real and measurable and actually quite consistent in scientific literature. You can ask large groups of women about their experiences of racial discrimination and then chart their birth outcomes and see the strong and lasting correlation between that discrimination and premature birth. You can measure certain stress hormone levels and notice that Black women with higher levels of this hormone are more likely to give birth prematurely. You can, as Collins has, compare red-lined neighborhoods in Chicago against a map of high rates of preterm birth and see that they are, in preliminary data, tightly associated with each other. Zip codes that have more racial and economic segregation—a marker for structural racism—have higher rates of premature birth for Black women.

The University of California, San Francisco, in particular, is doing groundbreaking work on this topic. A recent study led by Dr. Brittany D. Chambers and Dr. Monica McLemore found that three-quarters of pregnant Black women regularly experienced racism in their daily lives, such as at school, at work, at a store, or on the street. Chambers is now working on SOLARS, a study that aims to be the largest ever to focus solely on Black and Latina/x pregnant women: The research tracks self-reported experiences of racism and other chronic stressors like adverse childhood events, alongside stress hormone levels and other biomarkers in blood, urine, and saliva.

Crear-Perry has come to understand her own preterm birth very differently over the past two decades. Now an obstetrician-gynecologist and the president of the National Birth Equity Collaborative, an advocacy group, she is a leading voice arguing that researchers and health care providers should not cite "Black race" as a risk factor for preterm birth. Instead they should specify that "exposure to racism" is the problem. She can draw the lines directly between cause and effect in her own life.

Crear-Perry, a youthful, apple-cheeked woman with long braids and relentless energy, grew up in Grambling, Louisiana, a majority-Black town that is home to Grambling State University. Her mother is a pharmacist and her father an ophthalmologist; she remembers her childhood as sheltered and happy, surrounded by highly educated families who worked in academia or medicine. In 1989 she went to Princeton, where she remembers being called "colored" for the first time but also where she joined a supportive, tight-knit Black sorority, ran track, and excelled academically. In her last year of college, she gave birth to her daughter (past her due date) and married her boyfriend, a police officer. She got into medical school at Louisiana State University Shreveport. Things were going according to plan.

But something happened in medical school: Racism was pervasive. She had experienced discrimination previously, of course, but never like this: constant, virulent, frightening. She started to be afraid in a new way. She worried that she would not succeed not because she couldn't do the work but because some of the residents on her hospital rotations responsible for her evaluations clearly did not feel comfortable with a high-achieving, Princeton-educated Black woman.

This took many forms. Some incidents made her fear for her safety, such as when a white classmate writing up the shared lecture notes drew pictures of Black people swinging from trees like monkeys and strapped into electric chairs. And there were other, more common occurrences that made her fear for her professional future: She was often assumed to be a tech or an aide, not a doctor in training. She had lost that sense of freedom she felt in the community of her childhood, the agency she'd felt at Princeton, and was keenly aware that her future might rise or fall based on someone's false idea of her. There was also the fact that the LSU medical center, now called University Health, was called Confederate Memorial Medical Center until 1978.

"There are not many people who look like me who finished med school," Crear-Perry told me. "And so they were not accustomed to seeing me there. You can't sound too angry, too smart, too— You know, you have to fit in all the little boxes and make them feel okay. Which is nerve-racking. Every day. That's a scary place to live."

Crear-Perry draws a direct line between that constant fear and what happened next—a direct line between her first pregnancy, during which she felt safe, and her second pregnancy, during which she felt that she was in danger. In early June 1996 she was not even showing yet. When she started having abdominal pain, her first thought was that it was probably nothing. Ligament pain, Braxton-Hicks contractions, all those reassuringly normal things. She drove her four-year-old daughter to swim class and found that she couldn't get out of the car. She asked another parent to walk her daughter into class and drove herself to the doctor. She started vomiting and realized that this might be something to worry about after all. When they checked her at the hospital, she was already dilated and having strong, regular contractions. There was nothing anyone could do to stop her labor; the baby was coming that day.

In Crear-Perry's mind, she was miscarrying. The baby would be too immature to live. It was almost unheard-of for a 22-week baby to survive at any hospital anywhere at that time.

Meanwhile, her husband didn't have all the medical context that she did. He heard the doctor asking if they should intubate the baby at birth. He wanted everything done. Crear-Perry thought her son was going to die no matter what, and she wanted her husband to feel like they had tried their best. So the couple agreed to intubate.

Her son was born just a few hours later; amazingly, he cried briefly before doctors inserted the breathing tube. "I was, like, 'Look at him!'" Crear-Perry said, remembering that first surge of pride. But she was still sure he would die. And if he didn't, she worried about what his life would be like.

Incredibly, against impossibly long odds, her son, Carlos Jr., survived and came home after more than four months in the NICU. Nothing has been easy, for him or for his parents. Crear-Perry told me about the physical therapy, occupational therapy, feeding challenges, and the like that they have been through over the years. Her son has some developmental disabilities and hearing loss. These were her hopes for him: "I prayed: 'God, he might not be able to do all the cool things, but I want him to love and be loved.'" Now, she says, he does and he is.

Crear-Perry no longer blames herself, her skin color, for her son's early birth; she blames the environment she lived in, which made her feel fearful and powerless. She advocates a human rights approach to reproductive care. "In the genetics camp, they think it's simpler to look for a gene, because racism is too hard. To me, it's simpler to value everyone." This is not a matter of semantics. "It inspires a change in other people and not just telling Black folks what *they* should do differently, which is a big pivot," she said.

Slowly, this distinction is gaining traction, especially in some circles of researchers and in the media. But it is quite far from widely accepted and disseminated; the idea that skin color is the source of the risk is still dominant.

But the research is clear: Experiencing discrimination is bad for your health. Racist systems are bad for the health of the nation. We are dynamic, porous organisms. We are made of our mothers and fathers, but also the food we eat, the air we breathe. We are all shaped—both psychologically and physiologically—by our environments. Our cumulative life experience becomes embodied in us. This is true of everyone, but it has meaning and consequences that can be specific to Black women. In speaking to Black women who have had a preterm birth, asking them about racial discrimination, I was struck by how many of them used metaphors like "the air I breathe" or "the water I swim in" to describe the pervasiveness of the experience.

Another facet of this problem is that pregnancy necessitates interaction with health care systems, and Black women tend to experience discriminatory health care even in the best hospitals. That includes rougher handling during cervical exams, unanswered questions, and untreated symptoms. (Think of Serena Williams and the oft-told story of how she almost died after her C-section because her doctors didn't believe her when she said she had a blood clotting condition.) Research has shown that doctors tend to undertreat and underestimate both women's symptoms and Black people's symptoms. For Black women, that's double jeopardy.

In November 2017, I went to San Francisco for a three-day research symposium put on by UCSF's Preterm Birth Initiative (PTBI). The conference was called "Racism and Preterm Birth," a purposefully jarring title. The information presented at the symposium was devastating. I often found myself swallowing back tears, particularly listening to a presentation by the PTBI Community Advisory Board members, a group of women of color who have had premature births and who advise the UCSF researchers on the lived reality of the phenomenon.

Hope Williams is a special education assistant teacher and member of the Community Advisory Board. We met one evening on Treasure Island, a quiet hamlet in the San Francisco Bay populated by apartment buildings and old navy barracks. Hope looked younger than her thirty-eight years, except at certain moments, when a curtain of exhaustion descended over her face. We sat at a picnic table watching the sun set over the glittering San Francisco skyline and talked about the birth of Williams's little girl.

About nine years before Williams got pregnant with her second child, she discovered that she had high blood pressure. To get it under control, she lost weight, changed her diet, and started exercising more. She also started seeing a physician regularly, a doctor who was caring and attentive and who put her on a particular blood pressure medication for a short period of time. By the time Williams started planning to get pregnant again in 2014, she felt confident that she was in good health.

But, as is often the case for women who have had high blood pressure in the past, during pregnancy Williams's blood pressure rose again. She knew this was not ideal, but no one explained to her that this put her at risk of a preterm birth and that it put her own life at risk. She kept going back and forth to the hospital so that they could get her blood pressure down, only for it to rise again. (She had had to change insurance plans, and the physician who had been so helpful before was no longer in network.) She knew that the certain medication that she took years before had controlled her blood pressure quite well, but when she mentioned it multiple times, no one listened, instead giving her other medications that didn't seem to have any effect on her.

"I didn't understand what was going on," Williams remembered of the end of her pregnancy. She was sick and afraid; she couldn't get anyone to answer her questions but she could tell she was getting worse. "My hands were so swollen, I could barely hold a fork to feed myself," she said. She and her daughter's father had just thrown a baby shower; their friends and family were excited to meet the new baby. Williams wanted to be celebratory, too, but the truth was that she felt terrible and she worried that, despite the doctors' seeming lack of concern, something was really wrong.

About a week later, when she was 34 weeks along, she got up in the morning and felt her water break. So she called her daughter's father to tell him to meet her at the hospital. She put a towel between her legs, packed bags for the hospital. Then she made breakfast for her older daughter, dropped her off at school, and drove to the hospital.

She was in active labor already. "I didn't understand what this meant for my baby," Williams recalled. "I didn't even know what the NICU was." The birth happened fast, and her daughter came out crying. "I'm, like, 'Just give me my baby.' No. They take her and they rush her out. I'm like, 'Where are you taking my baby?' They didn't tell me. Finally, the nurse is like, 'They took her to the NICU.' I'm like, 'What is the NICU?'" Williams sent her daughter's father after their baby; she was too sick to go, incapacitated on heavy-duty medications to prevent a stroke, as her blood pressure continued to rise.

Finally, the doctor who had cared for Williams all those years before, who was no longer in her network, heard that Williams was sick and in the hospital. It was her day off, but she came by anyway. She listened to the whole story and reassured Williams that she wasn't going to die. She put her back on the blood pressure medication that had worked before. Within a day Williams's blood pressure was normal.

This story has a happy ending: Both Williams and her baby girl, who is now a spunky preschooler, are healthy. Williams joined PTBI's Community Advisory Board in the hopes that she could help other women avoid what had happened to her. But the fright of her daughter's birth still reverberates for her; she knows now that both her life and her daughter's life were at risk unnecessarily when her blood pressure kept spiking.

To think about what happened to Williams, start upstream: Her experiences with racism throughout her life may have caused her high blood pressure to begin with. The connection is well documented. Throughout her life she was treated differently, dismissively, aggressively, in both work and health care settings. Then, when the problem presented, if someone had just believed her and put her on the drug she knew had worked before, she might not have gone into preterm labor. At the bare minimum, she should have been informed of her risk of premature birth and told in advance where her baby would be taken after birth. Multiply her story by the hundreds of thousands—not all of them with happy endings—and you get a sense of the scale of this crisis.

These stories are finally getting out. There has been a spate of reporting on the epidemic of Black maternal and infant mortality—which are closely entwined—and their direct links to the experience of racism. Around the time of the UCSF symposium, it felt like a dam breaking: The March of Dimes, which is dedicated to preventing preterm birth, started focusing on the racial disparity as the most urgent facet of this problem. The New York Times Magazine put Black maternal and infant mortality on the cover; NPR's Priska Neely covered it in depth for weeks. Perhaps the volume of scientific evidence—and of human suffering—is finally too great for those in power to ignore: Democratic presidential candidates senators Kamala Harris, Elizabeth Warren, Kirsten Gillibrand, and Cory Booker have proposed various bills and plans that seek to address the issue.

Dr. James Collins, the Northwestern neonatologist who has been doing this research for decades, says he can feel it, too, an increased willingness to at least acknowledge that there are real and measurable health consequences to the experience of racism. "I remember when I first said it, it was very hard to say in a public venue," he said. "It's been a dramatic change. Dramatic."

Prematurity is a lens: It can illuminate. Prematurity is a canary in a coal mine; it can tell us that something isn't right. Prematurity can help us see what might otherwise be hard to understand or accept. Or the research on the connection between racism and prematurity can shed light on our own experience. (For some people, of course, this news is not news.) I notice a tendency toward activism in my fellow NICU parents—a desire to do anything we can to prevent premature birth from happening to other families. One thing we can do, *especially* those of us who are white, is listen to the people who have been doing this work all along and follow their lead. So I'll follow Crear-Perry: The problem is racism, not race. And it doesn't have to be this way.

16

What Prematurity Means in Mississippi

EVERY NOVEMBER, WHEN the March of Dimes releases its yearly prematurity reports, Dr. Charlene Collier braces herself for a fresh round of "Mississippi is the worst." It's true: Mississippi is the state with the highest rate of preterm birth in the nation—13.6 percent of births are early. This drives Mississippi's infant mortality rate, also the highest in the nation. Collier is not in denial about this. As an obstetrician-gynecologist, she sees her own patients give birth early, and as the director of the state's Perinatal Quality Collaborative (MSPQC) and a research consultant at the health department, she reviews and tallies the state's preterm births and infant deaths. Every day she tries to improve the health of women and their babies, and in Mississippi that means a lot of metaphorical walking uphill. About 20 percent of all Mississippians have incomes under the poverty line, the highest poverty rate of any state in the nation—another unfortunate superlative and a key risk factor for preterm birth.

What Collier really braces for, though, is the lack of nuance. Something bad is happening in Mississippi—too many babies born early, too many babies dying—but there's no sense of *why* this might be happening, in what populations, and, more importantly, what can be done about it. "The worst" makes an attention-grabbing headline, but it's harder to get people to pay attention to the details. There is a kind of voyeuristic inertia, a collective shrug.

There are layers upon layers to the high preterm birth rate in Mississippi: many women affected by poverty; many women affected by racial discrimination (the state is about 36 percent Black); and many women living in areas that do not have any obstetric providers. Nearly 50 percent of Mississippi counties are what the March of Dimes has termed "maternity care deserts." And there are plenty of women who are affected by several of these layers.

What this adds up to is that many Mississippian pregnant women are chronically, clinically stressed, hit hard by what public health experts call social determinants of health. That means factors like lack of safe housing, workplaces, and neighborhoods and limited access to education, health care, and nutritious food.

Here is just one example: Low-income women are more likely to have blue-collar jobs that require you to be on your feet all day. Collier says, for instance, that she has pregnant patients who do heavy lifting for forty hours a week. Their bosses don't care—don't *have* to care—that the women are pregnant, even though the NIH says that long working hours on your feet is a risk factor for preterm birth. In most states, including Mississippi, employers have no lawful obligation to give breaks or lighter physical work to pregnant women.

On a humid August morning in 2018, I met Collier at the Mississippi Department of Health in Jackson, where she spends three days a week as head of the PQC setting policy standards for obstetric care within the state, in addition to her clinical work seeing her patients and delivering babies at the University of Mississippi Medical Center.

Collier grew up in New Jersey and attended Brown University

for undergraduate and medical school. Then she got a master's degree in public health from Harvard, did her residency at Yale, and did a fellowship with the Robert Wood Johnson Clinical Scholars program. She planned to stay in Connecticut but moved to Jackson because her husband is from the area. She had always wanted to work with underserved populations, and once she started practicing in Mississippi, she could see how much work there was to be done in this state that does not have enough providers to go around. "We chose to stay here. He's the reason we came; I'm the reason we stayed," she said.

We sat in her office, which has a large window overlooking the medical center where she works. Her desk displays several photos of her husband and small boys. She has a friendly but intense and cerebral way about her, a habit of rapping her desk with her palm to emphasize a point. She looks like the kind of person who forgets to eat lunch. She communicates even the most frustrating information with disarming matter-of-factness. She swiveled her chair to turn on her computer and look at her most recent unpublished data on preterm birth in Mississippi.

These numbers fluctuate in small ways all the time, but it was and remains true that far too many Mississippians give birth early: The statistics that day confirmed the 13.6 percent average. If you break it down by race, Black Mississippians gave birth early 16 percent of the time, while white Mississippians gave birth early 11.3 percent of the time. That's a wide and problematic racial disparity, but Collier pointed out that Black mothers' rate had, for the moment, been holding steady, while the rate was increasing for white mothers. Collier didn't have conclusive data on why the white prematurity rate was going up. "My speculation? Opioids," she said. It makes sense: In the twenty-eight states that collect this data, the number of women giving birth while addicted to opioids has quadrupled since 1999. Vermont had the highest number of women who used opioids during pregnancy, at 4.8 percent. Misuse of or withdrawal from opioids can lead to premature birth.

The fact that Black women's rate was holding steady was a hopeful sign. Collier tentatively ascribed some of these modest gains—both an overall decline in extremely preterm birth and the holding steady of the Black rate—to the state-wide push to get doctors to use the only two medications we currently have to prevent preterm labor: progesterone shots (called 17P or Makena) and low-dose aspirin. These medications work only for particular groups: Weekly progesterone shots may be effective for pregnant women who previously have given birth early due to spontaneous labor; they lower the risk of a repeat early birth. And low-dose aspirin is indicated for pregnant women with high blood pressure, lowering risk of preeclampsia.

These medications are safe for the mother and baby, yet for a variety of logistical and financial reasons a surprisingly low number of women who might benefit from them actually get them.

Collier's last estimate was that fewer than 25 percent of the Mississippi women who need 17P get it. Progesterone is typically available only in its branded version, and it can be quite expensive—hundreds of dollars per shot—and needs to be given once a week for twenty weeks. Collier advocated for it to be covered by Medicaid and other major insurance plans without prior authorization, which can eat up precious weeks. Then she worked on accessibility: Since pregnant women in Mississippi have to drive long distances to see their providers, weekly visits to get a shot are unfeasible. So Collier successfully lobbied to make it permissible for pharmacies and home health organizations to administer the shots. Even so, she says, getting progesterone shots for just one woman can turn into a time-consuming nightmare. "First it's the insurance company, then it's the pharmacy, then it's the pharmacy's distributor, and who exactly has to make a phone call to the distributor to ship it to this pharmacy."

But then there are the stories that keep her going: A patient came to her after having a 17-week miscarriage and then a 24-week baby who died in the NICU. Collier felt deeply for this woman and made sure she did everything possible to get her to term, including weekly progesterone shots. The woman gave birth to a healthy baby at 37 weeks.

Unlike progesterone, low-dose aspirin is cheap and easy to get. But doctors have to tell their patients they need to be on it and why. Some pregnant women are leery of taking any medication; some providers don't know the evidence in favor of it or lack the time to explain it. But Collier is dogged about getting the word out and then making the word into policy.

Much of what she does involves setting up policies that educate hospitals and providers and then incentivizing them (or requiring them) to do the right thing, even if the right thing takes a little more time and effort. This is how the state cut down on nonmedically indicated early elective births. (That's a C-section or induction before 39 weeks for no medical reason, just because the patient or the doctor prefers it.) The health department got together with insurers and told hospitals that if someone was delivered early for no reason, the hospital wouldn't get reimbursed for that delivery. It worked. Collier prefers a carrot to a stick, but she'll use either one.

"I love health policies. Because then you have to do it. You have to do it regardless of the patient's color, creed, background," she said. "That is what I believe in now more than anything. I have faith in the influence of policy on practice." Collier laughed sadly.

It is relentless, this work. It sounds like a game of Whac-A-Mole played blindfolded. Then there is the added complication that some of Mississippi's state leaders are actively, though perhaps unintentionally, fostering an environment that leads to preterm birth.

In vast swaths of the state there are no obstetricians whatsoever. These maternity care deserts are not inevitable or naturally occurring; they may be the result of political choices. Mississippi is one of the states that has declined the federal government's offer of more money to extend Medicaid to additional adults. At first this seems to have little bearing on pregnancy—pregnant people already qualify for Medicaid at much higher incomes than the general population—until you realize a few things: For one, people who go into pregnancy healthy are more likely to have full-term babies. So giving everyone access to affordable health care makes financial (and ethical) sense—a recent Georgetown analysis found that states that expanded Medicare had healthier mothers and infants.

Then there is the fact that rural hospitals are shutting down because they can't afford to stay open: Too many of their patients are uninsured or underinsured; either these people don't go to the doctor at all, or when they do, they can't pay their bills and the providers and hospitals don't get paid. This exodus of doctors and medical facilities from rural areas means women have to search for a provider and often drive long distances both for prenatal care and to give birth, which can be especially dangerous when the pregnancy is high-risk or the labor is premature. An infusion of Medicaid dollars would help. Rural hospitals have fared much better in states that have expanded Medicaid, because when more patients are insured, more bills get paid, and more hospitals can afford to stay open.

In fact, if the Mississippi state government *did* accept the Medicaid expansion, it would provide insurance for about 300,000 Mississippians—more than 10 percent of the state's population. These are the people who make too much to qualify for Medicaid, but not enough to afford to buy their own insurance. At the moment, for a Mississippi family of four, the household income must be less than $569 per month (about $6,800 yearly) for parents to get Medicaid coverage.

Contraception is one way of preventing premature birth, not just in the obvious sense that it prevents pregnancy, but because when pregnancies are spaced at least eighteen months apart, they are more likely to be full term. But when state legislatures like Mississippi's restrict abortion in every way they can—there is a single Planned Parenthood clinic in the state—they are not only restricting access to abortion, they are also limiting access to contraception. It's not as if these communities have another gynecologist down the street.

This interplay of policy and culture has deep, real-life implications for women's and babies' health. A recent study from Ibis Reproductive Health showed that the more restrictions on abortion in a given state, the higher the maternal and infant mortality. This certainly holds true in Mississippi, which is one of the five states with the most restrictions on abortion and is the state with the highest infant mortality rate in the nation.

This is not just because some women who wanted an abortion were not able to get one and ended up having unhealthy pregnancies. It's even bigger than that: According to the Ibis report, states that passed more restrictions on abortion also tended to pass fewer laws protecting the health of pregnant women and children, including statutes that prevent premature birth, like requiring bosses to make reasonable changes to job duties to accommodate pregnancy and mandating that workplaces be smoke-free. Restrictions on abortion aren't just restrictions on that one medical procedure: They seem to foster and flow from a culture that doesn't prioritize the health of women and children.

Contrast all this to Vermont, which, like Mississippi, has large expanses of rural, agricultural land, and which also has a considerable opioid problem. It is true that Vermont is overwhelmingly white, and so most women there are not contending with racism. But it's also worth noting that Vermont's leaders have different priorities from those in Mississippi: Vermont has expanded Medicaid as well as its state insurance exchanges, and only 3 percent of people in the state are uninsured (as opposed to about 18 percent in Mississippi)—one of the lowest rates in the nation. It has sufficient maternity care. It has no restrictions on abortion. And it has the lowest preterm birth rate in the nation and one of the lowest infant mortality rates. The research suggests that these things are all connected, inextricable.

Wengora Thompson, Mississippi's maternal/child health and government affairs director for March of Dimes, works closely with Collier. "I've seen states where you have an entire bureau that's devoted to what she does," Thompson says of Collier.

You could say the same of Thompson herself. The March of Dimes' Mississippi headquarters is in a neat brick bungalow at the end of a cul-de-sac in a mostly residential area. When I arrived, I thought I must be in the wrong place. I had driven down a quiet street lined by lawns, and it seemed as though there was no one around. But, sure enough, Suite B was occupied by Thompson, who was working alone that morning.

Thompson grew up in Mississippi and then moved to Ohio for grad school and worked at the Ohio Department of Health, collecting data and working on infant mortality initiatives. She noticed that many more resources were dedicated to maternal and infant health in Ohio than in her home state.

In her current role, Thompson allocates March of Dimes funds—about $45,000 per year—to programs that target preterm birth, and then she coordinates those initiatives and the data that result. Most recently that included a pilot program at the University of Mississippi: At routine baby visits to the pediatrician, office staff asked the baby's mothers if they smoked, if they were depressed, if they were using contraceptives, and if they were taking folic acid.

The idea was that the babies are likely to be getting regular checkups because they are covered under Medicaid, while their moms might be uninsured because Medicaid can be revoked when you are no longer pregnant. Doing outreach to the moms through the pediatrician was a kind of end run around the fact that these women might be otherwise unreachable by health care providers. Depending on each woman's answers to these questions, the office staff could then refer the moms to services they might want: smoking cessation programs and contraceptive or mental health providers. Obviously that's good for a woman's health, which is an end unto itself, but it's also a way to prevent premature birth before the hypothetical next pregnancy even starts.

Thompson has deep frustrations with Mississippi, starting with some doctors who don't provide care based on the latest evidence—like not offering progesterone and low-dose aspirin—and going all the way up to the state government and its insistence on not expanding Medicaid. This has become a partisan issue—Republicans against, Democrats for—because it was a part of President Barack

Obama's Affordable Care Act. But the March of Dimes is a nonpartisan organization; it does not advocate for one party or the other. It just looks at policies and asks one question: How does this affect preterm birth and other negative pregnancy outcomes?

Thompson sees a layered and deeply entrenched problem in Mississippi: high poverty intersecting with a long and ongoing history of racial discrimination intersecting with state leadership that seems, for the most part, uninterested in the problem of preterm birth. Thompson says this lack of attention to the issue is shortsighted to say the least. "It's not just a save-our-babies problem," she said. "Prematurity is an economic issue. It's taxing on both our health care and education systems, because the research has shown prematurity impacts the whole life course."

Progesterone shots, for instance, might be expensive. But they aren't as expensive as NICU care, follow-up care, and all the loss of productivity that flows from premature birth, first on the part of the parents and later on the part of the grown child, who may end up with health or developmental issues. And this vicious cycle only deepens poverty.

Collier, for her part, embraces the big-picture-ness of premature birth and the way preventing it all comes back to caring about each other—not in a vague, feel-good way but in ways that literally hold the key to a well-functioning society.

"If, as a whole society, we're not taking care of pregnant women, imagine being a poor Black woman in Mississippi. You're going to get treated poorly in many areas of your life. You're not getting loved on. And I have no doubt that very poor white women are not getting loved and respected either," she said.

We have poor birth outcomes, she argues, at least in large part because of high chronic stress and the environments in which pregnancies happen in this country. You might not be able to find an obstetrician at all, or you might only be able to find one who treats you disrespectfully; you might have to hide your pregnancy to get a job; you might have to work full-time stacking boxes until you go into labor; you might have to think about finding (and affording!) day care the minute you see the plus sign on the stick, because the very second that baby is out, you're going to have to hobble back to work.

This is especially true in Mississippi, yes; it can be especially true for marginalized communities, yes; but it is also true nationally. If you want to gauge the health of a given community, you could do worse than to just look at the preterm birth rate. If it's high, there's something wrong. Collectively, our bodies are telling us something.

Collier says that when the many layers of this issue overwhelm her, she just puts her head down and focuses on doing what she can, like getting progesterone and low-dose aspirin to all the women who can benefit from it. And she will keep seeing her own patients. She'll deliver their babies. She'll keep taking care of Mississippi as best she can.

17

Group Prenatal Care and the Power of Community

WHAT WOULD IT look like to remake prenatal care into something that would inoculate—even in a small, temporary way—against serious stress? It would look like this: a circle of chairs in a Bronx hospital, and a midwife who listens more than she speaks. It's called group prenatal care, and it is exactly what it sounds like.

This particular hospital, Bronx-Lebanon (now called BronxCare), generally has a 12 percent prematurity rate. Among group prenatal care participants in 2018, that rate was 4 percent. That is better than the premature birth rate in Scandinavia, which has the lowest rates in the world. (It is probably not a coincidence that those countries also rank as the happiest in the world, with robust social programs that protect citizens from the anxieties that plague Americans.)

For an intervention that can have such a dramatic effect, group prenatal care is astonishingly simple, cost-effective, and low-tech. It simply involves gathering pregnant women together for a support group and open discussion along with prenatal care. Women receiving group prenatal care still get all the same standard checks everyone else does—fetal heart monitoring, belly measurement, weight and blood pressure checks, and so on—but they also spend about an hour sitting together: a small circle of women who are due around the same time, asking questions, giving opinions, and getting information. And this, just this, might cut their risk of preterm birth in half.

CenteringPregnancy, run by the Centering Healthcare Institute, is the oldest, largest, and most data-validated of all group prenatal care models, although March of Dimes and the insurance company UnitedHealthcare have recently launched a similar program, as have researchers at Yale. United is one of several insurance companies that believes the model is cost-effective, in some cases offering enhanced reimbursement (a.k.a. more money per patient) to providers who use it. That's because they think it saves them money in the long run. In a South Carolina study, the use of Centering slashed NICU admissions from 12 percent to 3.5 percent, saving thousands of dollars.

Although the evidence in clinical trials has been mixed, a recent study by Yale and Vanderbilt universities found that patients who attended five or more group prenatal care appointments cut their risk of a preterm birth by 68 percent. Centering's own internal, unpublished data show that their patients had a 6.7 percent rate of premature birth in 2017, while the average American rate is 10 percent and climbing. There may be some selection bias in that the women who choose group care and follow through with it might be different from those who do not, but that number is all the more impressive when you consider that the population served by Centering is higher risk than the average: About half their patients are low income. The evidence is compelling enough that the U.S. military now offers group prenatal care for its pregnant service members.

The power of this model seems to rest on the group being a temporary stress buffer, an empowering source of support. And it's not just an individual-level intervention: It restructures and reworks the way that care itself is provided, making it more likely that patients will be heard, seen, listened to. Perhaps for these reasons it has so far been shown to be most effective with groups that have the most chronic stress and highest likelihood of being mistreated by the medical system: low-income women and women of color. But it can benefit anyone.

Dr. Monica McLemore, professor of nursing at UCSF, thinks that group prenatal care works on several levels: It is a more transparent, open way of providing care and giving information, it provides peer-to-peer learning, and it minimizes time spent alone with a provider.

"I like group care because it hones right in on the problem, which is how we are transferring health information during pregnancy, how we are treating pregnant people, how we are thinking about them both as a parent and as a member of a community. It's transformed. It's not expecting somebody to be a passive patient behind a door with a gown on." These factors, she said, add up to a kind of secret sauce: The women coming to group prenatal care might be less likely to be mistreated or dismissed by their provider since there is safety and power in numbers. And they may feel freer to be their authentic selves.

To see for myself how these groups work, I sat in on a series of CenteringPregnancy sessions at Bronx-Lebanon Hospital. The group was made up of Black and Latina women, while the midwife who ran the group was white; this is a not-uncommon dynamic, though one that is worth noting.

It was a cool Wednesday in April, and three women—Grace, Myesha, and Kathy—trickled into the small room on the fourteenth floor of the hospital. On this particular day only three of the six group members were able to make it. (I am using first names throughout this chapter for patient privacy.) They had been meeting every few weeks for the last three months, and this was their sixth session together; they were now between 30 and 33 weeks pregnant. The three greeted each other warmly, a patter of catching up, comparing aches and pains. Late-afternoon sun slanted in the window and illuminated the boxy brick landscape that surrounded the hospital. Leah Halliburton, the certified nurse midwife who facilitated the groups, passed around a basket of snacks, asked everyone how they were feeling, and mentioned that she had ordered breast pumps for everyone. Today, she said, they were going to talk about pain management for labor.

One by one, each of the three women weighed herself and recorded the number. Then they each took their turn on a blue recliner by the window. They pulled up their shirts and let Leah measure their bellies. She told them the number and recorded it, remarking to each woman that everything looked good. Then Leah squirted some blue gel on a tissue, rubbed it on the woman's belly, and used a handheld fetal heart monitor to listen to the baby. As if by unspoken agreement, there was a lull in the group's conversation as Leah slowly moved the wand around each swelling belly, until a rapid gallop filled the small room, the baby's heartbeat momentarily the only sound.

We scooched our chairs into a circle. On one side of the room was a series of prints showing the stages of pregnancy, a pregnant torso in cross section, the baby getting bigger and more baby-like as the headless woman's internal organs got more squished in each proceeding poster. On another wall there was a Georgia O'Keeffe poppy print, and on another a black-and-white print of a Hindu sculpture depicting the goddess Kali giving birth to the universe.

Myesha, a bubbly, high-cheekboned young woman who was

pregnant with her first baby, hadn't been able to sleep, so she'd been up all night watching YouTube videos of births, and now she was, very understandably, freaked out. "They were pushing on this woman's stomach to make the placenta come out," she recounted, shuddering. "And they had snipped her . . ."

"An episiotomy?" asked Leah.

"Yeah," said Myesha.

"Our practice almost never does those," replied Leah.

Myesha seemed a bit reassured. But what about the placenta? she wondered out loud. How hard was it to push out, after all the work you'd already done getting the baby out?

Grace, a calm, quiet woman with long braids who has an older son, recalled it as not a big deal. "It came out in one push, like jelly," she said.

"Did you look at yours?" Leah asked Grace.

"No!" said Grace, making a face.

A comfortable pause as Leah passed out cups of water and completed some paperwork, then Grace had a question: Did she *have* to have her baby put on her chest, skin-to-skin, the moment he emerged? Could the doctor take the baby away and wash him first? Leah explained that the latest research suggested that babies shouldn't be bathed right away, because the goop all over them after a vaginal birth contains good bacteria that will make them healthier in the long run. But, Leah said, moms certainly don't *have* to do skin-to-skin right away. "When you are in early labor, try to remember to let the provider know if you think you won't want immediate skin-to-skin," she said, reiterating that that was a totally acceptable option. We discussed the logistics of this, and exactly how much goop we were talking about, and the current thought on delayed cord clamping. The group hadn't even officially started yet and I had already learned more about the physical realities of giving birth than I did in six months of obstetrician's visits.

Leah suggested that we turn our booklets to a page that listed pain management strategies: rhythmic movement and breathing, massage, warm baths, epidurals, and other medications. She gave everyone a few minutes to read and then asked if they had thoughts.

Myesha remarked that she thought rocking back and forth would help her deal with the contractions. But she also wanted an epidural, for sure, even though her mother kept telling her *she* had had six babies without meds. Grace said that she had an epidural with her son, and it wore off before pushing. This time she wanted to try to do it unmedicated. Myesha was horrified to learn the epidural could wear off early. Kathy, a warm, sweet-faced woman with a sheet of dark hair under a Yankees cap, said that after having epidurals with her two older girls, this time she wanted to try to do a water birth.

Leah listened and affirmed that these were all good options, answered questions, and explained that everyone needs some pain relief strategies that weren't an epidural, because you can't always get an epidural right away and, yes, sometimes it wears off. She moderated and offered information but didn't dominate or direct. She guided the women to answer each other's questions with their own insight. She also shared her own experience in giving birth to her son and asked that I share mine as well. In Centering, everyone speaks personally because the act of sharing or withholding can create a power imbalance. In traditional prenatal care, the provider wears a professional mask, while the patient is exposed, literally and figuratively.

"I don't want to give you a lecture about it, but here are some good visuals," Leah said, passing around a sheet that explained the stages of labor (early, active, transition, pushing) and a plastic tablet with progressively larger holes in it to show how wide the cervix opens over the course of labor. (Ten centimeters is . . . very large.) Leah explained that, as the cervix widens, it also thins, finally allowing the baby's head to pass through, and that's why you can't push until you're fully dilated. She showed this by encircling her forehead with her hands. "The baby's head can't move when there is still cervix under it," she explained. "It would be like pushing against a brick wall."

Myesha wanted to know if it was true that you should only push when you are having a contraction, because her mom had told her it wasn't true. "That is mostly true," said Leah, "because the contractions are what help you push the baby out. But by the time your mom was on her sixth baby, she knew what she was doing and they probably let her do whatever she wanted!"

They were really digging into the nature of labor pain now: how bad, exactly, it would be, what pushing was like. Myesha was cradling her belly with her hands. Her joking and bluster had fallen away.

"I'm scared," she said, a catch in her voice.

"It's okay to be scared; that's part of it," said Leah.

Grace gazed at her from across the room. "I think you're more nervous than scared," she said.

Grace added that when she was in labor with her son, she felt like the pain was bad but manageable, even when her epidural wore off. When it was time to push, she said, smiling, "It was just like the worst constipation. Like day three of constipation, when you decide, 'Okay! I've *got* to do something about this.'"

Then Grace said that the worst pain she ever experienced, actually, was her miscarriage two years earlier. She recounted starting to spot a little, being worried, and then suddenly the bleeding accelerating, like a faucet. She went to the ER. "I was screaming," she said, and started to cry. Myesha replied that last year she had had an ectopic pregnancy, a condition in which the fertilized egg implants in a fallopian tube rather than the uterus, which can be life-threatening for the mother. Doctors had to end the pregnancy, which would never have been viable. She had to spend her recovery on the maternity floor, where she was surrounded by blissed-out women with brand-new babies. "I was so depressed," she said, and started to cry, too.

"They should not have put you there," said Leah. "I'm so sorry that that happened to you."

The women took a moment to let all that settle, and then the conversation continued to flow. Who was their chosen support person for labor? Were their (all male) partners feeling oriented to what was going to happen in childbirth? (Not really.) Circumcision: yes or no?

It was freewheeling and personal and supportive and poignant and funny and very, very informative. It was medical information *and* good conversation. By the time the two-hour session was over, we had talked about bodily fluids, mothers and siblings, thoughts on foreskin, miscarriage, weight, and leaky breasts. And while preterm birth was not at the top of the women's minds, this visit, and the cumulative effects of the previous visits, had likely cut their risk of an early delivery.

Leah, who is the Centering coordinator at Bronx-Lebanon, believes that group prenatal care can mitigate the risk of preterm birth because she has seen it herself. Leah says that the model's effectiveness makes sense to her both intuitively and medically. In particular, she says, for women who are not high-risk for a particular medical reason, Centering seems to mitigate the chronic stress that can lead to an unexplained early birth. "Because of what we know now about chronic stress, toxic stress, institutional racism, and the roles that those might play in preterm birth, it seems reasonable through my experience and what I've read that this would be an effective intervention for preterm birth," she said.

Leah also just prefers the model as a way to provide care. It is more personal; it allows her to be effective, giving her patients what they need. "Working in the traditional clinic, the time crunch is unbelievable. They expect you to see a patient every ten to fifteen minutes all day long. We have so many patients with complex medical issues as well as psychosocial issues, and it's just impossible to get through all of their issues in ten minutes," she explained. "You leave feeling like you didn't provide the best care you could have and you have patients who are upset because they've been waiting for a long time. It's just this constant stress ball.

"Centering allows you to get past the surface things and talk about a lot more," she continued. "And even with the more simple questions, it's good to have that group knowledge. In a lot of my groups there's a mix of first-time moms and experienced moms. I love the things that experienced moms throw out, like suggestions for what they did for their nausea or their backache, because sometimes it's better than the information we give them. That's one of the great things about that shared space."

The group members prefer it as well. Kathy noted that at a one-on-one appointment she might be too embarrassed to ask all her questions. "If I feel like it's a dumb question, I won't ask it," she said. "But here, I don't feel that way. Here, I just ask." Grace said she really likes the mix of giving and getting. "The more we talk, the more I remember what I want to ask," she said. "You can ask your questions and give your opinions."

It was immediately obvious to me why the group prenatal care model might work. It felt amazing—*amazing*—to sit there in a circle, even as a nonpregnant observer, to see how free everyone felt to speak, to offer opinions, to ask questions, no matter how big or small. I felt the relief in the room, the relief of putting down those heavy bags of worry, the things you can't stop thinking about late at night. I remembered frantically googling low PAPP-A in my office and trying in vain to get my doctor to talk to me in detail about what I had found. I thought about how different it might have been to bring it up here, to hear medical information, yes, but also simply to have my fears validated. To put that fear down. I don't know if it would have made a difference in outcome, but it would have made a difference to me.

Group care can also be customized to speak directly to the people it serves. At South Dakota Urban Indian Health in Sioux Falls, providers have adapted the model specifically for their community, mostly made up of people from the Lakota, Dakota, and Sioux Nations. The program, called Baby Steps, is focused on storytelling, which is culturally resonant for the women who attend: They share previous birth stories, stories about mothers and grandmothers. Even for a topic as prosaic as car seat safety, the women are encouraged to talk about their own experiences while also getting information about the correct way to use one, and of course they also get the standard prenatal checks. During the discussions, they make traditional crafts like baby moccasins, cradleboards, and leather amulets. Nationwide, about 11 percent of all Native American births are premature. Among the women who have participated in Baby Steps so far, there has been only one early birth.

In the Bronx a month or so later, the women were gathered again: a little more tired, a little more pregnant, and very hot. The temperature had suddenly soared into the eighties even though it was only the first week of May. "I'm just trying to make it to my baby shower," said Myesha. Kathy walked in, saw Myesha, and broke into a huge grin. "Hiiiiiiiiii!" exclaimed Kathy as the two women hugged.

Leah took Myesha to a private room for a routine vaginal swab, and Kathy went off to a scheduled ultrasound. In the meantime Destiny and Natalie, two women who had missed the last session, had arrived and splayed in their chairs, marveling at how hot it was. Leah and Myesha came back and Leah asked Destiny to sit in the big blue armchair so she could measure her belly and the baby's heartbeat. Destiny said she wasn't sure if the baby was head down yet, so Leah gently palpitated Destiny's stomach.

"Yup," she said. "Head down."

"You just touched the head?" Destiny said in awe.

"It's easy on you," said Leah.

Destiny felt her lower abdomen with an amazed look on her face. "So I'm touching his head right now?" she asked. Leah nodded.

Kathy burst back in, crying with happiness, holding up a strip of four black-and-white photos. "I was bawling in there," she said. "He's perfect. He's got a lot of hair." She showed the photos to the other women and everyone made swooning noises. She squinted at one of the photos and asked Leah what it showed. "Uh, this is some . . . testicles!" said Leah.

Leah had an orange balloon in her hands and she was stretching the mouth open to fit a Ping-Pong ball inside. Then she blew up the balloon but didn't tie it; the Ping-Pong ball blocked the mouth and kept the balloon inflated. "Okay," she said. "This is supposed to simulate labor." She giggled.

Myesha started laughing. "That's supposed to be our private parts?"

Leah almost dissolved into laughter but held it together. "Okay, so the contractions start from up here, the top of the uterus, and this"—she pointed to the mouth of the balloon—"is your cervix." She started squeezing the balloon from the top, the pink Ping-Pong ball fell down to the mouth the balloon, and the mouth started to stretch open.

"That looks crazy," said one of the women.

"I'm about to pass out," said another.

"You are going to pop it!"

Leah kept squeezing. "This is a long process!" she said, and showed how the rubber ring of the balloon's mouth was starting to stretch open and more of the Ping-Pong ball was visible. The women were in hysterics.

"It's a girl!" someone yelped.

"Can you not just pull it out?" said someone else.

"That's not how labor works!" said Leah. "You have to keep going!"

And then: *Pop!* Everyone, including Leah, screamed, and then screamed with laughter as the balloon burst and the Ping-Pong ball went flying. "Okay, well, that was a problem with the uterus," said Leah. "That's not what happens."

Everyone collected themselves and then they talked more about the mechanics of labor and pain management: what they wanted, what they feared. At the end, Leah passed out board books—their choice of *The Very Hungry Caterpillar* or *Goodnight Moon*. Myesha gently held the book to her belly.

As the weeks went by, I started to look forward to group, just to see how each of the women was doing, to see their bellies get bigger, to hear about their baby showers and their older kids.

At the next session two weeks later, Myesha, Grace, Kathy, and

Leah were midway through a conversation about C-sections—why they are sometimes medically necessary and how to avoid one if possible—when Sasha, a strikingly pretty, statuesque woman with a leonine mane of curly golden hair and purple lipstick, banged in, clearly upset. She collapsed on the chair; the other women greeted her warmly. Sasha started to cry.

She recounted a formidable list of frustrations: She was staying at a city shelter for domestic violence survivors, which had a 9:00 p.m. curfew, which meant that last night, when she got hungry, they wouldn't let her out to go to the store. The shelter administrator had been rude, dismissive, wouldn't even give Sasha a bottle of water when she asked for it. Her employer (she works as a CNA) was giving her a hard time about scheduling maternity leave, telling her conflicting information about the paperwork she needed, sending her on one wild-goose chase after another. She was also in the midst of taking her ex to court. Sasha was 36 weeks pregnant, living alone, feeling alone. "I feel like a juggler," she said tearily, miming juggling with her hands and feet at the same time. "I am in the dark."

The air in the room changed; everyone quieted, as if to give space to the hugeness of Sasha's despair and anger. No one told her to calm down. Leah put her hand on Sasha's shoulder and told her how good it was that she was there.

"You'll get your own place," said Myesha. "You have a plan."

"It's okay, Sasha," said Grace quietly. "Think of the baby. Leave your ex in the dark and you be in the light."

"This is your safe space," said Kathy.

When Sasha was ready, Leah measured Sasha's belly, checked her blood pressure and her baby's heart rate. The other women, meanwhile, were in problem-solving mode, planning to bring Sasha gift certificates and baby supplies next time, since she wasn't having a shower. When Sasha bemoaned the ten dollars she had spent on cab fare to get to the hospital, Myesha told her that Medicaid would pay for cabs to prenatal appointments if she called in advance.

Sasha's situation is a good example of the kind of overwhelming stress that puts someone at high risk for preterm birth. No one thinks that group prenatal care can solve the challenges Sasha is up against; it would be condescending and simplistic even to suggest it. When she left the group that day, she still had to deal with the shelter, her work, her ex. Those systems were still stacked against her. But it's possible that group prenatal care helped her have a healthier pregnancy, for herself and her baby. If she had gone to a traditional OB visit that day, it's hard to imagine it going as well. Would a provider who has to see a new patient every ten minutes have been able to respond well to such a complex and emotional encounter? Would Sasha have repressed her anger and frustration? Would she have let it out? I could imagine the same scenario ending in a doctor asking Sasha to calm down, to lower her voice; I could imagine it being a moment when another system failed her. Instead, that day Sasha was heard. She vented. She received kind words and some pragmatic suggestions. And she got prenatal care. It was not enough, but it was something. The circle held.

About two months later, on an oppressively humid August Wednesday, Leah was arranging wraps, fruit, Caesar salad, and brownies on the table in the meeting room. On the whiteboard she had written "Congratulations!!!" and listed the names of all the babies who had been born to the women in the group: all healthy, all full-term. Destiny was the first to walk in, her long, beaded braids swinging, pushing a stroller. Her baby boy was sleeping, his head tilted to one side, leaning on his puffy little cheeks. She and Leah hugged, and Leah exclaimed over the perfectness of the baby.

"I miss coming here!" Destiny said, helping herself to some lunch, fanning away the stickiness of the city. Kathy came in next, with her two young daughters in tow and pushing a stroller with a two-month, chunky baby boy. He gurgled and burbled, looking around with wide eyes. Then Myesha came in with her sleeping son, and suddenly the room was full of strollers and brand-new humans, and everyone was squealing over the babies, exclaiming about how good everyone looked, asking how everyone was feeling. How were the babies sleeping? How were they eating? Leah held each one of them in turn, marveling at them, giving Myesha's boy a bottle. The women recounted birth stories, and it turned out that Myesha was the only one to have to have a C-section, an outcome she had been dreading. "Look at what you put Mommy through," she said gently to her son. They were all tired, hot, happy to see each other.

One intervention is never going to be enough to truly address the various problems that lead to preterm birth. Dr. Joia Crear-Perry, the obstetrician who had her son at 22 weeks and now works on reproductive health equity, says she thinks group prenatal care is great, but she worries that it gets oversold as the answer to every problem. "I do think it's really good for patients who need social connectedness," she said. "The hard part is that we like shortcuts—proxies for figuring out things like race or income." Her point is that group prenatal care might work really well, particularly for certain patients, but the effectiveness of group care doesn't let us off the hook for addressing these big problems that lead to people needing a support group in the first place.

In Mississippi, Centering groups just started to be available in the autumn of 2018 at Jackson-Hinds Comprehensive Health Center. Dr. Collier thinks this is an important step, and she would like to see the model expand to more locations in her state.

"What about it is magical?" she asked. "Well, it centers on a person, helps them to tune into emotions, relieve stress, feel loved and important, take time for themselves. A pregnant woman is taken into a space where they are loved, respected, and empowered around their health. And *somehow* they end up with fewer preterm births."

Part VII

The Invisibles

Breaking the Silence

18

The Hidden Trauma
of Prematurity

PREMATURITY CAN BRING up big bioethical dilemmas, but the drama of most premature birth is small and private, mostly unseen. The vast majority of premature babies are successfully treated in the cloistered NICU and go home. About 80 percent of premature births occur at 32 weeks or later, with an average survival rate greater than 95 percent. But for these babies and for their parents, the consequences can reverberate for years.

There are about 3 million American children under the age of ten who were born between 32 and 36 weeks. They are the invisible majority, not the one-pound miracle babies you see on the local news or in your Facebook feed. These later-born babies have better outcomes overall than babies born earlier. Their time in the NICU is shorter, and less likely to cause them major harm—and yet, it is far from ideal and can result in delays and disabilities. Their parents' trauma is also real and yet sometimes feels as though it must be hedged when spoken of, mixed with gratitude at all times. Almost everyone who has spent any time in a NICU knows it could have been worse.

But it's all how you look at it: Babies who are born at moderate-to-late preterm gestations have many more challenges overall than full-term babies. A recent study found that these babies are between

3 and 9 times more likely to have delays or impairment in speech, cognition, and motor skills even *after* adjusting for socioeconomic factors. They are more likely to struggle with friendships, play, attention, and behavior, and this can lead to increased prevalence of learning disabilities once they are ready for school.

Despite this, as we've seen, those born at 32 weeks and older may not qualify for follow-up programs or any sort of Early Intervention services. Their parents may have to scramble to get them the help they need—or receive mixed messages about their child's level of risk.

Liz Rodriguez Sowek was staying overnight with her mom in the Bronx when she woke up in the middle of the night bleeding profusely; she was 31 weeks pregnant. She went by ambulance to nearby Lincoln Hospital, where she was diagnosed with a partial placental abruption, in which part of the placenta separates from the uterine wall. The bleeding subsided, but she had to stay on bed rest for monitoring; it was too dangerous to be transferred. More hemorrhaging was a distinct possibility and an early birth was all but inevitable. The goal was to make 34 weeks.

So Liz remained on hospital bed rest while her wife, Emily Sowek, had to go back to Albany, where the couple lives, to take care of their pets and to go to work. Emily came back and forth as much as she could. But this meant that later the following week, when doctors realized that Liz was in active labor, she was alone. Emily jumped in the car and started the two-hour drive. But meanwhile the baby had to be born.

"It [the C-section] happened so quickly, everyone was running around like chickens with their heads cut off. And all I could think of was 'I'm 32 weeks and five days. I didn't make it to 34.' The doctor was telling me 'You did a great job. You did everything I asked you to do and more. You did nothing wrong.'" Liz started weeping as she spoke.

I knew how she felt. It is a profound grief over the idea—however false—that your very first act as a mother was to fail your child. To fail them fundamentally, bodily. It is still ragged for me, too, and maybe it always will be.

"They took her out and I couldn't hear anything for a good minute. It was the longest minute of my life. Then I heard a sound like a weak kitten. The doctor brought me my daughter; she looked like a little burrito. Then all I could think of was how beautiful she was and how sad I was for my wife, who couldn't be there."

Liz's daughter, Noelle, was 4 pounds, 10 ounces. She had supplemental oxygen via nasal cannula for a day and needed to be tube fed for several weeks. She stayed in the NICU for twenty-two days. There is no question that, from a neonatology perspective, that is a pretty good scenario. But for Noelle's mothers, those twenty-two days were a trial. After Liz's discharge, Emily had to go back to work. Liz was on maternity leave and stayed with her mother to be close to the hospital; every morning she got up at dawn and took the bus to the hospital, where she stayed all day.

"I'd sit outside the Isolette and just talk to her all day long. I would put my hand in the Isolette and stroke her cheek or her head and tell her how much we love her and how we're so happy to have her. We'd do skin-to-skin [kangaroo care] and I'd sing lullabies. I was so nervous with the wires. I'd ask to have her put back every time I felt like crying and then I would leave the room and bawl."

Liz didn't want Noelle to feel or hear her crying.

"When I'd leave the hospital, it was already dark. And I'd be waiting for the bus in the cold, seeing people pushing baby strollers or pregnant women. My heart just hurt. If people were smoking cigarettes, I'd cover my nose and hold my breath, because I still felt pregnant. I cried on the bus. I cried walking from the bus to the hospital; I cried at CVS getting my medications. One time the head of neonatology was walking past, and I was outside of the NICU in the hallway in the fetal position on the floor, just bawling my eyes out. I felt so alone."

Sometimes it seems we are living in a moment that gives space for increasingly inclusive and realistic stories about what it means to give birth and become a mother or a parent. Chrissy Teigen posts pictures of herself wearing mesh postpartum underwear (still looking like Chrissy Teigen) and talks about her postpartum depression. I know the degrees of my friends' perineal tears and how breastfeeding went for them, or didn't. It's healthy for these realities to come out of the dark and private spaces where women have had to keep them for so long. It's resulted in a bit more leeway for narratives about parenthood that are not instant bliss. Still, we have a long way to go.

Experiences like the one Liz described are extremely common, though they are still rarely spoken of—are hard to speak of. And these experiences can take a lasting toll on mental health: Mothers of premature babies are at higher risk for postpartum depression, for instance. And parents can have symptoms that meet diagnostic criteria for PTSD as a result of a NICU stay.

The other day I heard a midwife being interviewed on the radio, saying that birth is something a woman's body knows how to do, not a medical event. It made me flinch. I thought: *Not my body.* Not millions of other bodies, either. The emphasis on "natural" birth that is meant to be empowering can be painful for those of us who needed every unnatural intervention to get our babies (and/or ourselves) out alive. Even the difficulties that are most often talked about—difficult labor, how to get babies to eat and sleep—seem to exist in a different universe than the NICU. The conversations just don't translate: Did you breast- or formula feed? *I tube fed.* Do you still check every five minutes to see if she's breathing? Who wants to be the one who brings up apnea of prematurity or SIDS data at the new parent meet-up? My neighborhood has parent groups that are organized by birth month. I didn't know if I should go to the group for Mira's adjusted or actual age. I was weirdly ashamed of my own confusion and I never got up the courage to go to either one. I also didn't want to be the bearer of dark information, and I couldn't imagine how I would participate in "normal" conversations about my baby.

For parents of later-born preemies, it doesn't help that NICU clinicians often say things that are meant to be reassuring but can feel minimizing and confusing. In NICU jargon a baby who doesn't need much medical support is called a feeder-and-grower. The nickname is based in a medical reality—those babies mostly just need time and calories—but it doesn't do much to acknowledge the very real risks these babies do face or the grief that parents might be going through.

For instance, at discharge, Liz asked for Noelle's medical records. A clinician said sure, they'd get her the records, *but it wasn't like Noelle had had a rough stay.*

"I get it," Liz said. "Noelle didn't have to be intubated or her heart didn't fully stop. I get it. I get it. But when I thought about it in retrospect, I thought, 'How dare she?'"

Noelle is now four months old. Since she's been home, she's suffered with a severe form of reflux common in preemies, which Liz struggled to get her pediatrician to diagnose. And she was rehospitalized in the pediatric intensive care unit for respiratory syncytial virus (RSV), a common respiratory virus that can make premature babies critically ill.

"I'm constantly nervous that something else is going to go wrong," Liz said. That feeling is hard to shake, because even when a child is fortunately healthy, there are lingering effects of prematurity that can be hard to predict. It's a difficult balance for any parent to strike: just vigilant enough, but not too vigilant.

Kristen Mascia, a journalist in the Bay Area, had a similar experience. Her son, Sam, was born at 32 weeks in 2017 to her and her husband, Tom. Mascia had previously had cervical cancer, which necessitated two biopsies that shortened her cervix, putting her at risk for preterm birth. She started having contractions at 28 weeks and her water broke at the start of her 32nd week. Sam was born weighing just over four pounds and was admitted to the NICU, where he stayed for twenty-eight days.

Mascia started reading, vacuuming up every bit of information she could find about what this meant for Sam. What she learned about his risks for developmental, sensory, and learning delays terrified her.

"I remember everybody saying, 'The fact that he reached 32 weeks is great,'" said Mascia. "It was cold comfort, because at the end of the day, statistics still suggested that he could very well go on to develop problems. And I *was* grateful. But it's hard to feel lucky when you're in a room full of tiny babies in incubators with beeping monitors and flashing lights. I just felt so guilty that he was in there. I felt shame around not having fulfilled my first fundamental duty as a mother to protect my child."

Like Liz, the only time Mascia wept was when she spoke of this guilt.

"To see this poor baby. And again, compared to some of the other babies, he seemed like a moose. And knowing what those other parents were going through . . . there was guilt associated with that as well. Like, 'How dare *I* feel this way?'"

Mascia struggled to hold all of these truths at once: Her child was lucky; her child was in intensive care and would face heightened risks for many developmental and learning problems; she was lucky; she was grieving.

"There was dissonance on several levels, not only between what the doctors were saying and what I knew to be true based on my own research, but also just emotional dissonance between what we were dealing with, with friends and family, who often had no experience of preemies and didn't really know what to say. It was hard to reconcile."

When Sam came home, Mascia couldn't sleep for vigilance, terrified that he was going to stop breathing, which had happened in the NICU. She also worried about autism, knowing that premature kids are more likely to develop it. She watched and watched for eye contact, social smiles, language development, all of which came in time but seemed to take forever. This is the thing about adjusted age: Preemies act like newborns for a really long time. Mascia knew that Sam couldn't be expected to smile until he reached the appropriate adjusted age, not his actual age. But it can be easy to lose faith in that idea, to panic. After the NICU, sometimes panic seems like a very appropriate reaction.

"The first month of his life at home was just not a joyful time," says Mascia. "Everyone wants to celebrate. They want to see the baby. They want to welcome you home. It's difficult to respond in a way that feels normal. You don't really even realize how many expectations you have about what bringing a baby home is going to be like or what childbirth is going to be like. And everything about this experience dashes those ideas."

Many parents of premature infants say that they felt alone, like a failure. That they had absolutely no reference points for understanding what was happening to them and their children. They were dropped into a hidden world full of critically sick babies and then emerged, weeks or months later, expected to move on as if nothing had happened, to celebrate and leave their guilt and fear behind. From the outside it looks simple: your baby was too small; now your baby has grown. There are not a lot of cultural reference points for what prematurity really means to families, to communities. Cabbage Patch preemies, for instance: If you were alive in the 1980s you might remember them. They're just extra-small, bald versions of the regular dolls. They definitely don't come with nasogastric tubes or heart monitors or prescriptions for physical therapy. But when these experiences stay invisible, millions of parents struggle to get the help they need, for themselves and their children.

19

Grown Preemies
Speak for Themselves

SINCE THE MOMENT egg warmers were adapted for babies, the smallest infants have served as blank slates for our projections. We project onto them our anxieties about everything from the nature of motherhood to what makes life worth living and who should get to decide. Debates about quality of life often have a problematic power dynamic involved: the abled discussing the merits of disabled life. But our imaginations are limited, and no one is an expert in someone else's quality of life.

Around 1973, Dr. Saroj Saigal, a neonatologist at the then brand-new NICU at McMaster University in Hamilton, Ontario, noticed that parents of very premature babies tended to ask a question she couldn't answer: Will my child wish they had not been saved? At this time, the first generation of intensive care survivors was being discharged. These children had never survived in large numbers before, so no one had a sense of how they would do long term. This made it hard to counsel parents when they asked what the chances were that, if their baby survived, their lives might somehow be worse than death.

Saigal was worried about the babies and the parents, with whom she often became close. She wanted to know how they were doing.

She was worried about this brand-new medicine she was practicing: What if she was doing more harm than good? What if these children's lives were painful, unfulfilled, unhappy? She wanted to be able to answer the parents' questions honestly.

So she became one of the first researchers to follow extremely premature babies over the long term. Her research on a group of 179 people born at under 1,000 grams between 1977 and 1982, amazingly, is still ongoing as the participants enter their forties. She has traced their lives all this time.

By the time they were three years old, one-quarter had a significant disability of some sort; about 10 percent were blind—rates of blindness have dropped considerably since then—and 15 percent had some level of cerebral palsy. About half of them experienced developmental delays. Saigal and her colleagues noticed, though, that some of the children's challenges subsided over time as they caught up and gained skills. Other issues emerged over time—like learning and attention differences—as the kids progressed in school.

At eight years old, for instance, the formerly premature kids were much more likely than control-group counterparts to repeat a grade or use special education services. But, on average, these difficulties were not so severe that they affected the qualities you might say really matter: Prematurity had no average impact on the kids' relationships with parents and friends or their abilities to enjoy hobbies, music, school clubs, and extracurriculars. At adolescence, the results were similar: more learning and attention difficulties, although social and emotional development were generally unaffected. (Later, there would be some evidence of increased risk of depression and anxiety.) And with each interval of time, some difficulties abated, especially in terms of general health and hospitalizations.

Thirteen of the children were so severely disabled or ill with lung disease that they did not survive childhood; that intense suffering is part of the truth, too. But the overall picture that emerged, as the years rolled on, was one of resilience: The formerly premature children experienced more hardships than their term-born, control-group counterparts, but their lives were really not very different from anyone else's, even among those with a disability. When they did report significant quality-of-life problems, it was often around discrimination and bullying because of a disability, not deficits caused by the disability itself.

Still, the conclusion of Saigal's studies each time reflected the bottom-line results of standardized testing of motor, social, and cognitive skills, and educational measures: People born at extremely low birth weights fared worse than the control group by many measures of health and functioning.

But Saigal was starting to feel that this kind of measurement, this comparison, was not telling the whole story.

"In the long run, I got tired of this number game," she told me. "So many with cerebral palsy, so many with blindness, so many [with] deafness, you keep adding them. It reminded me of *Alice in Wonderland*: What's one and one and one and one and one? We didn't have the true meaning for an individual child." Dr. Annie Janvier, who trained under Saigal at McMaster, would echo this idea years later.

In her book, *Preemie Voices: Young Men and Women Born Very Prematurely Describe Their Lives, Challenges and Achievements*, Saigal mused on this question. She wondered if the scores on these standardized tests said anything significant about a child's life or if they just provided a snapshot of their abilities at a certain moment in comparison to a "typical" child. She acknowledged the usefulness of the scores in comparing hospital to hospital and for researchers in speaking a common language and publishing their results. "But how do we capture the overall functioning of an individual child in a way that is meaningful for the child and the family, taking their personal values and preferences into consideration? We had no idea how to do this as it had never been done before in children or adolescents," she wrote.

If quality of life is inherently subjective, then only each individual can really answer the question of whether or not they find life worth living. So Saigal decided to abandon the standardized tests and ask the children, who were at the time teenagers, what they thought of their own lives. "I said, 'I don't care what their IQ is. I don't care what scores they get in the language and motor tests. I want to see how they are functioning in everyday life.'"

Saigal had difficulty finding funding for this new kind of study, which used standard gamble questions to quantify how highly individuals valued their own health and state of functioning. Many funders felt that asking the children themselves was problematic—that they somehow couldn't be trusted to know how they felt. Some suggested asking their parents about the children's quality of life instead. But Saigal was determined: Who could know better about an individual's subjective experience of their own state of health and well-being than the individual themselves?

The results of this new kind of inquiry caused a stir: Extreme prematurity and being born under 1,000 grams (2.2 pounds) had only a marginal impact on how the participants themselves rated their own quality of life in adolescence.

Based on their answers to the study questions, each participant received a score that ranged from zero (death would be preferable) to one (utterly perfect health). Overall, the former preemies rated their quality of life as only slightly lower than the control group. For instance, 71 percent of former preemies considered their quality of life to be better than .95 (very close to perfect), while 73 percent of controls felt the same. In fact, many of those with disabilities rated themselves as enjoying very good and in some cases near-perfect well-being. Conversely, some control group individuals without any health issues rated their quality of life as quite low.

This scientifically demonstrates something intuitive: How an individual experiences their own life is not obvious from the outside. (Saigal's paper was the first to make the assertion that people who have a disability do not necessarily consider themselves to have a lower quality of life than other people, a finding that has now been replicated several times by other researchers.)

Some people found this very hard to believe. Much of the pushback Saigal got revolved around the idea that the teenagers in her study must not have understood the questions or that they were in denial. Saigal thought it much more likely that the participants simply felt they had better lives than other people expected they would—including her. The results of the study had surprised her, too.

In her book, Saigal recounted a story about one memorable individual: This teenager had severe cerebral palsy and used a wheelchair. He rated his quality of life quite highly. When the interviewer responded with surprise, the boy explained that he was doing well in school, he was close with his family, and that overall he felt happy. What else could he say?

"In hindsight, the problems reported in the earlier years painted an overly pessimistic picture for the future—because, at least in part, the comparisons were always made to 'normal' function and much less to how these young people were doing in their own terms," Saigal wrote.

When Saigal expanded this study to include all the parents of the former preemies, asking them how they felt about their children's quality of life, it turned out that parents and children mostly agreed. This says something about family cohesion, perhaps, but it also shows that parents are, in fact, good surrogate decision makers for their babies—that they can rely on their own values, beliefs, and understandings, knowing that their children are likely to generally agree with them about what constitutes a good life as they grow up.

Saigal repeated the quality-of-life study when her participants reached adulthood. At that point, the quality of life margin between the former preemies and the term-born control group narrowed even more: There was virtually no difference in self-defined quality of life.

Conversely, when Saigal and her colleagues asked neonatal doctors and nurses to assess the quality of life of those with various disabilities, the clinicians were much more likely to rate those states as being worse than death. It's not totally clear why this is; it may have something to do with medical professionals valuing conventional academic and social success, or it may have to do with their sense of heavy responsibility, knowing all the things that can go wrong or being haunted by the worst-case scenarios.

Meanwhile, parents of disabled children and the children themselves mostly adapted to their realities, showing tremendous resilience. They got on with it; they did what they had to do. And in many cases the parents came to realize that their children's state of being, whatever it was, wasn't as difficult as they might have feared. Love probably has a lot to do with this.

This reminds me of something that Janvier said about what her own daughter taught her and how her personal experience changed how she speaks to parents in the NICU. "What I didn't know that I realize now is that the majority of people adapt to the worst things," she said. "What I wish someone had said to me was that one day it's going to get better. One day you'll be okay. Nobody told me that. And no, life is not sending what you deserve or what you can take. It's shit, but it's going to get better.

"Parents of preterm children see life differently afterwards," she added. "They're more vulnerable, but they're also stronger. They learn things about themselves they wouldn't have learned without that. It's hard for those who have very disabled kids, but there are also some positives about it that nobody speaks about."

Likewise, Saigal does not want to sugarcoat how hard it can be to parent a premature child. But she does think it's important to convey what her research says about how families adapt and how good quality of life can be, even for children who end up with a serious disability, rather than just conveying what doctors feel about these scenarios, which is clearly more pessimistic. Saigal feels that it is not only humane but also medically accurate to offer hope, even when the prognosis is uncertain. And she also feels comfortable now answering that question: Children do not grow up and wish they had not been saved.

For Saigal's book, she asked her study participants to write letters about their own lives so they could speak for themselves. At the time, the participants were in their mid-thirties, and their voices are as varied as you would expect of any large group of people linked only by the circumstances of their birth. For many, the experience of having been premature becomes a kind of origin myth, something they've heard about all their lives. For some, this translates into a feeling of gratitude, of being one of the lucky ones, having narrowly escaped death at birth. Some hold it as an integral part of their identity, while others are connected to it in a mostly abstract way. Some of them have not thought about being premature much at all.

Some note that prematurity was not something they experienced themselves but something that was reflected back to them in the emotion of their parents. One woman remembered speaking on-stage at a first-grade event and seeing her mother run out of the auditorium in tears, overwhelmed with emotion for her daughter's accomplishments.

Rishi Kapur, who was born at 27 weeks and weighed 1 pound, 14 ounces, and is now a physician, wrote of this in an arresting way.

"What is most challenging about being born premature is my inability to connect with my early life. It is like waking up in a world where everyone else has experienced war except yourself and there is no evidence of the aftermath."

April Laramey, who was born at 27 weeks and 2 pounds, 3 ounces, has mild cerebral palsy, graduated from college cum laude, and now works for the Canadian government in Ottawa. Writing in Saigal's book, she put it this way: "The stories I have heard all my life about my birth seem like mythology to me—like the origin story of some-one else's superhero tale. . . . To parents of premature babies who may face similar difficulties in their lives I'm afraid there is no per-fect advice. Every child is different and unique, including those with disabilities."

The stories are moving because they are both extraordinary and reassuringly familiar. These people are all exactly my age and were born at around Mira's birth weight. They write of being rushed from one hospital to another. They were ventilated and operated on; they were said to be dying. Now they pose with their own young children on their laps, with their arms slung around their spouses, smiling in the sun. They wear jerseys from their favor-ite sports teams. They note things like how long it took them to walk, whether or not they were bullied at school because of their limp or learning disability. They feel they owe their lives to various forces: to the doctors and nurses, to luck, to science, to God. They like country music, rock music, playing the flute. They are doctors and government employees, librarians and carpenters. They like camping and gardening and bowling and drinking beer with their friends. They are blind and sighted; they use wheelchairs or canes; they are able-bodied. One avid racquetballer's cerebral palsy manifests in a very slight limp. Another has more significant cerebral palsy that results in intellectual disability and difficulty walking—she swam in the Special Olympics. One has fetal alcohol syndrome and struggled with a childhood spent in foster care. One woman who has hydrocephalus is also an avid scuba diver.

When you're pregnant, it is natural to imagine your child as a perfect blank slate, unbound by any limitations. They could be anything: an Olympian, an astronaut. It seems cruel to take even an ounce of potential away from a baby. An early arrival into the world changes our bodies and minds in ways big and small, even on a cellular level. It is embedded in our psyches and in our bodies in both literal and metaphoric ways. But the world writes on all of us, whether we are born early or on time. The meaning we make of our lives belongs to each of us.

Epilogue

*Live on, survive, for the earth gives forth wonders. It may swallow
your heart, but the wonders keep on coming.*

—SALMAN RUSHDIE, *THE GROUND BENEATH HER FEET*

MIRA AND HER friend A., both three years old, are playing at a
neighborhood café that has a dress-up area and a ball pit and a whole
host of other diversions, and A.'s mom and I are chatting, drinking
tea, and watching them. The girls tussle with giant foam blocks like
tiny WWE wrestlers, they pull on tutus and superhero capes, they
pass a doll back and forth and zoom around on little wagons. A.'s
mom and I haltingly share the stories of our girls' births with each
other: A. was born with a heart condition and almost died as a
baby; she was saved by a successful heart surgery.

"Look at them; they have no idea," A.'s mom says, gesturing to
our girls, who are engrossed in telling each other what to do.

Born in another time, born right now in another place, neither of
them would be here. Their lives, their wiggly little bodies, are man-
ifestations of the work and belief and struggle of countless people.
And there is so much more work that needs to be done. Maybe these
girls will be the ones to do it.

When I first saw Mira, it hurt me to look at her body, to see her
exposure, her malnourishment. Everything about her seemed to be
an indictment of my own failure to protect her: How little she re-
sembled a baby, a comfortable baby with little fat rolls, a baby you

could hold and feed. She was skin and bones; she aroused the same kind of pity in me as a hurt animal, that kind of reflexive struggle to survive, the blind insistence of life choosing itself.

Now, when I find myself in a NICU and I see babies Mira's size or smaller, I can see their delicate, filigreed beauty. Premature babies are travelers from another world, beings from the in-between. Caring for them is like touching some essence, something you aren't supposed to see: the terrible beauty of a human being built, gorgeous and unknowable.

Sometimes Amol observes that Mira and I still interact like we're one body, how there still seems to be a cord connecting us. I like that he says this, that he notices this, because that bodily connection has been so hard-won, so painfully interrupted and painstakingly reestablished.

About a month after we got home from the NICU, Mira suddenly started refusing the bottle, and through trial and error we realized that she would only drink if we placed her on her side on the couch next to us and fed her while sitting next to her. There was no holding her while she drank the milk that I had pumped. She often seemed overwhelmed. She would sleep on me but not snuggle when she was awake, although she was soothed by being in the carrier. She was not a lovey-dovey baby, although she was happy, sweet, giggly, and often delighted by our cat, books, the bath. Maybe I'm projecting, but she seemed a little wary. The last time she was snug in my body, things went very badly for her. She seemed disinclined to give it another shot.

But somehow, at around eighteen months, she started to snuggle, recreationally. To throw her arms around my neck, to cuddle on the couch while she drank her bottle of (thank God, cow's) milk.

It felt like forgiveness—her of me, me of me—however ridiculous that may sound. But it also proved to me how resilient and elastic those bonds are. How, if you keep trying, if you show your child care and love—whether you have gestated your child fully, partially, or not at all—your child will love you back. Our relationship is something we built together, over time, in daily acts of caring for each other.

I wanted to understand what happened to us; I wanted to understand premature birth. I have a much richer understanding now, informed by other families' experiences, by clinicians' expertise, by public health experts' alarm. So, what can premature birth teach us about being human? There's no one answer, just more questions. Maybe that's appropriate.

What does it mean to gestate a child? What does it mean to mother a child who has not finished gestating? How should we protect and care for the very smallest and most vulnerable among us? How can we move toward becoming a society in which everyone feels safe enough to have a chance at a healthy pregnancy? How can we stop pretending that childbirth and child-rearing is something individuals can do alone?

Prematurity teaches us, forces us, to live with unanswerable questions—to lean into the mysteries. It is living with doubt and fear and grief and nevertheless moving forward, because we must, in our imperfect, human ways.

Gestation is a marvel, no matter how it is accomplished. But gestation isn't the real magic. Love is the magic. Love is the only magic.

Acknowledgments

THE GESTATION OF this book has been a group project. I am filled with gratitude for the people who made it possible, and deeply indebted to all of them.

Scores of you took time out of your busy lives to share your personal stories with me, to lend your professional expertise, or both. Others were early readers of the work, giving support and feedback without which I would have been lost. I can never repay any of you, but I can offer my deepest thanks. This book belongs to you, too.

To the parents who opened up about your children's births, and in some cases their deaths: You shared some of the most difficult days of your lives to make others feel less alone. Not all of your stories ended up in the book, but they are here in spirit. Thank you for your generosity: Marcie Haley, Erin Alexander, Kavita Williams, Tyrese Coleman, Kesha Evans, Abbi Roberson, Maybelin Carmona, Princy Abraham, Brittany Lothe, Rosa Martinez, Lametria Burks, Dr. Jen Gunter, Chrissy and Jordan Hutchinson, Nancy Bautista, Laura and Graham Shullman, Alexis Kropp-Kwon and Young Kwon, Aleshia Jones, Hope Williams, Alejandra Sanchez, Liz Rodriguez Sowek, Kristen Mascia, Traci Messenger, and Dr. Deborah Campbell.

And thank you to those who shared about your own early births: Beth Allen and Martha Lott.

Thank you to the clinicians, researchers, and health care professionals who took the time to answer my endless questions and show me how you do what you do: Robin Bisgaard, Dr. Monica McLemore, Dr. Brittany Chambers, Dr. Elizabeth Rogers, Dr. Larry Rand, Linda Franck, Dr. Dawn Gano and the team at UCSF, Dr. Edward Bell, Dr. Steven McElroy, Dr. Matthew Rysavy, Dr. Jonathan Klein, Dr. Tarah Colaizy, and the team at the University of Iowa. And to Dr. Edmund LaGamma, and the team at Maria Fareri Children's Hospital.

Thank you to Dr. John D. Lantos for being incredibly open and generous with your time and expertise.

Thank you to Dr. Anjali Chelliah, Dr. Bhaktharaj P. Chelliah, Dr. Jill Parekh, Dr. Brownsyne Tucker Edmonds, Dr. James Helm, Frances McCarthy, Dr. Jeffrey Horbar, Micah Sandager, Dr. Elliott Main, Dr. David K. Stevenson, Dr. Teresa Janevic, Dr. Neil Marlow, Felice Sklamberg, Dr. Joseph Kaempf, Dr. Brian Carter, Dr. Alan Fleischman, Dr. Mark Mercurio, Dr. Carl H. Backes, Dr. James Collins, Dr. Marlyse Haward, and Dr. Arthur James.

Thank you, Dr. Jasmin O. Chapman, Marvin Dale Jr., Kimberly Lott-Jones, and the team at Jackson Hinds Comprehensive Health Center. Thank you to Dr. Charlene Collier and Wengora Thompson—for taking the time to speak to me and for the important work that you do.

Thank you, Tanya Munroe, Leah Halliburton, Shelia Reed, and the team at CenteringPregnancy. Thank you, Dr. G. Brent Whitton, Amy Sudduth, Terrie Roberts, and the developmental team at Willis-Knighton.

At the March of Dimes, thank you to James Soohoo, Katie Sellers, and Cynthia Pellegrini. At General Electric, thank you to Karen Starr and Steve Falk.

Thank you to all those who shared knowledge and insight on the history of treating premature babies: Dr. Jeffrey P. Baker, Dr. Håkan

Sundell, Cheryl Major, Donna Staed, Dr. Lawrence Gartner, Corey Reese, Anne Drorbaugh, and Dawn Raffel.

Deep and special thanks to Dr. Heidelise Als, Dr. Annie Janvier, Dr. Joia Crear-Perry, Dr. Elvira Parravicini, Dr. Saroj Saigal, Dr. Maria Delivoria-Papadopoulos, and Dr. Mildred Stahlman. Your work is the heart of this book.

Thank you to Dr. Christy Cummings, who gave the manuscript an expert, empathetic read.

With profound gratitude to the providers who cared for me and Mira: Dr. Martha Caprio, Dr. Pradeep Mally, Dr. Brina Maldonado, and, of course, every single one of the nurses. Thanks to Dr. Melanie Wilson-Taylor, Mira's wonderful, careful, calm pediatrician. Thanks to Pam Berger, who has kept me as sane as anyone could.

Thank you to the friends and colleagues who read the book in its various incarnations and gave invaluable feedback. In particular: Yaran Noti, who is a stellar editor and thinker and an even better friend. Thanks to Michelle Rome Moser, Maria Luisa Tucker, Melody Schreiber, Emily Barone, Suzanne Cope, Elizabeth Gold, Laura House, Lisa Selin Davis, and Orli Van Mourick.

Thanks to everyone at Harper who lent their considerable talents to this book, in particular: Kate D'Esmond, Christina Polizoto, Milan Bozic, Katie O'Callaghan, and Alicia Tan.

The prologue of this book started out as an essay I wrote when Mira was about a year old. There's only one person who has been shepherding this material all along, and that is my insightful and thoughtful agent, Jonah Straus. He was the very first person to believe that my jumble of thoughts could be a book, and that I could write it. His commitment to the project has meant the world to me.

I am unbelievably fortunate to have gotten to work alongside Gail Winston. Gail's sharpness, generosity, and ability to engage deeply with the material both emotionally and intellectually makes her a dream editor. I am so grateful for her honesty and the care she has taken in making this book become the best possible version of itself.

Thank you for the endless love, guidance, and support: Jenny Hellman, Colette Eastwood, Erin Eastwood, the Cochrane family, and Jyotsna and Anil Mhatre. Thanks to my parents, whom I miss every single day: Phyllis Eastwood and Mario DiGregorio. When you taught me to give and receive love and to find happiness in books, you gave me everything I would ever need.

And thank you to Amol and Mira, for letting me write about us, and for being my people. I love you.

Notes

PROLOGUE: ONE BIRTH

1   a strong, strange: John D. Lantos, *The Lazarus Case: Life-and-Death Issues in Neonatal Intensive Care* (Baltimore: Johns Hopkins University Press, 2001), 28–29. Reprinted with permission of Johns Hopkins University Press.

8   actually mixed: J. G. Anderson et al., "Racial and Ethnic Disparities in Preterm Infant Mortality and Severe Morbidity: A Population-Based Study," *Neonatology* 113, no. 1 (December 2017): 44–54, https://doi.org/10.1159/000480536.

8   not at all conclusive: Teresa Janevic et al., "Association of Race/Ethnicity with Very Preterm Neonatal Morbidities," *JAMA Pediatrics* 172, no. 11 (November 2018): 1061–69, https://doi:10.1001/jamapediatrics.2018.2029.

8   inferior NICU care: Jochen Profit et al., "Racial/Ethnic Disparity in NICU Quality of Care Delivery," *Pediatrics* 140, no. 3 (September 2017): 1–10, https://doi.org/10.1542/peds.2017-0918.

9   necessitates delivery: American College of Obstetricians and Gynecologists, "ACOG Practice Bulletin No. 116: Management of Intrapartum Fetal Heart Rate Tracings," *Obstetrics & Gynecology* 116, no. 5 (November 2010): 1232–40.

10   I found out later: NYU Langone Medical System records for the author.

15   one-third of a cup of blood: Tom Lissauer et al., *Neonatology at a Glance* (West Sussex, UK: John Wiley & Sons, 2016), 81.

16   Newborns evolved: University of Oxford, "Babies Don't Just Look Cute, Scientists Find," University of Oxford News and Events, June 6, 2016, http://www.ox.ac.uk/news/2016-06-06-babies-don%E2%80%99t-just-look-cute-scientists-find.

17   only a few ridges and gyrations: Heidelise Als, "NIDCAP-Promise to Protect the Preterm Brain," Foundational Lecture, NIDCAP Federation International, July 4, 2018.

21   at risk for post-traumatic stress disorder: Debra S. Lefkowitz, Chiara Baxt, and Jacquelyn R. Evans, "Prevalence and Correlates of Posttraumatic Stress and Postpartum Depression in Parents of Infants in the Neonatal Intensive Care Unit (NICU)," *Journal of Clinical Psychology in Medical Settings* 17, no. 3 (September 2010): 230–37, https://doi.org/10.1007/s10880-010-9202-7.

23   which premature babies suffer from more: Barbara M. Ostfeld et al., "Prematurity and Sudden Unexpected Infant Deaths in the United States," *Pediatrics* 140, no. 1 (July 2017): 1–8, https://doi.org/10.1542/peds.2016-3334.

26   1.6 percent of babies: National Center for Health Statistics, final natality data. Retrieved March 23, 2019, from www.marchofdimes.org/peristats.

26   about 0.5 percent: UCSF Children's Hospital at UCSF Medical Center, Intensive Care Staff Manual "Very Low and Extremely Low Birthweight Infants," 2004.

26   increased her risk: Erik A Jensen et al., "Adverse Effects of Small for Gestational Age Differ by Gestational Week Among Very Preterm Infants," *Archives of Disease in Childhood, Fetal and Neonatal Edition* 104, no. 2 (May 2018): F192–98, http://dx.doi.org/10.1136/archdischild-2017-314171.

26 almost a half million mothers: J. A. Martin et al., "Births: Final Data for 2017," *National Vital Statistics Reports* 67 no. 8 (2018), accessed at https://www.cdc.gov /nchs/data/nvsr/nvsr67/nvsr67_08-508.pdf.

1: WHAT HAPPENED?

32 8 percent chance: March of Dimes Perinatal Data Center, "2018 Premature Birth Report Card: New York State," accessed March 22, 2019, https://www.marchof dimes.org/reportcard.

32 29 times more likely: National Down Syndrome Society, accessed March 23, 2019, https://www.ndss.org/about-down-syndrome/down-syndrome/. Thirty-five-year-old women have a 1 in 350 (or 0.2 percent) chance of conceiving a child with Down syndrome.

32 how to prevent it: Janet M. Bronstein, *Preterm Birth in the United States* (Switzerland: Springer International Publishing, 2016), 29.

32 not a single disease or condition: Ibid.

32 the leading cause of newborn death: "The Global Problem of Premature Birth," March of Dimes, accessed March 22, 2019, https://www.marchofdimes.org /mission/the-global-problem-of-premature-birth.aspx.

33 leading cause: "Causes of Child Mortality 2017," World Health Organization, accessed March 23, 2019, https://www.who.int/gho/child_health/mortality/causes/en/.

33 developmental problems: "Long-term Health Effects of Premature Birth," March of Dimes, accessed March 23, 2019, https://www.marchofdimes.org/complications /long-term-health-effects-of-premature-birth.aspx.

33 $26 billion: "The Impact of Premature Birth on Society," March of Dimes, accessed March 23, 2019, https://www.marchofdimes.org/mission/the-economic -and-societal-costs.aspx.

33 10.02 percent: B. E. Hamilton et al., "Births: Provisional Data for 2018," Vital Statistics Rapid Release, no. 7 (2019), https://www.cdc.gov/nchs/data/nvsr/nvsr6 7/nvsr67_08-508.pdf.

33 almost 400,000 premature babies: Ibid.

33 the industrialized world: Maggie Fox, "Preterm Birth Rates Have Increased in the U.S.," *NBC News*, June 30, 2017, https://www.nbcnews.com/health/health -news/preterm-birth-rates-have-increased-u-s-n778576.

33 15 million babies: "Preterm Birth," World Health Organization, accessed March 23, 2019, https://www.who.int/news-room/fact-sheets/detail/preterm-birth.

34 infant mortality: David Johnson, "American Babies Are Less Likely to Survive Their First Year Than Babies in Other Rich Countries," *Time*, January 9, 2018, http://time.com/5090112/infant-mortality-rate-usa/.

34 "are going up": "2018 Premature Birth Report Cards," March of Dimes, accessed March 23, 2019, https://www.marchofdimes.org/mission/prematurity-report card.aspx.

34 decrease in stillbirths: John D. Lantos and Diane S. Lauderdale, *Preterm Babies, Fetal Patients, and Childbearing Choices* (Cambridge: MIT Press, 2015), 35–39.

35 37 and 42 weeks: "Preterm Birth," World Health Organization, accessed March 23, 2019, https://www.who.int/news-room/fact-sheets/detail/preterm-birth.

35 most people conceive around week 2: "Due Date Calculator," March of Dimes, accessed March 23, 2019, https://www.marchofdimes.org/pregnancy/calculating -your-due-date.aspx.

35 or rupture of membranes: "What Causes Preterm Labor and Birth?" National In-

stitutes of Health's (NIH's) Eunice Kennedy Shriver National Institute of Child Health and Human Development (NICHD), accessed March 23, 2019, https://www.nichd.nih.gov/health/topics/preterm/conditioninfo/causes.

35   about two-thirds: Tom Lissauer et al., *Neonatology at a Glance* (West Sussex, UK: John Wiley & Sons, 2016), 81.

35   or both: "What Causes Preterm Labor and Birth?," National Institutes of Health's (NIH's) Eunice Kennedy Shriver National Institute of Child Health and Human Development (NICHD), accessed March 23, 2019, https://www.nichd.nih.gov/health/topics/preterm/conditioninfo/causes.

35   known medical explanation: "Preterm Labor and Premature Birth: Are You at Risk?" March of Dimes, accessed March 23, 2019, https://www.marchofdimes.org/complications/preterm-labor-and-premature-birth-are-you-at-risk.aspx.

35   preeclampsia: "Preeclampsia," March of Dimes, accessed on March 23, 2019, https://www.marchofdimes.org/complications/preeclampsia.aspx.

36   working unexpectedly: Terry K. Morgan, "Role of the Placenta in Preterm Birth: A Review," *American Journal of Perinatology* 33, no. 3 (2016): 258–66, https://doi.org/10.1055/s-0035-1570379.

36   irregularly shaped uterus: "What Are the Risk Factors for Preterm Labor and Birth?" National Institutes of Health's (NIH's) Eunice Kennedy Shriver National Institute of Child Health and Human Development (NICHD), accessed on March 23, 2019, https://www.nichd.nih.gov/health/topics/preterm/conditioninfo/who_risk.

36   against premature birth: All information on risk factors for preterm birth in ibid.

36   early again: Ibid.

37   as much progress: Janet M. Bronstein, *Preterm Birth in the United States* (Switzerland: Springer International, 2016), 15.

37   progesterone: "Progesterone Treatment to Help Prevent Preterm Birth," March of Dimes, accessed on March 23, 2019, https://www.marchofdimes.org/complications/progesterone-treatment-to-help-prevent-premature-birth.aspx.

37   low-dose aspirin: "Preeclampsia: Key Points," March of Dimes, accessed March 23, 2019, https://www.marchofdimes.org/complications/preeclampsia.aspx.

37   used a blood test: Laura L. Jelliffe-Pawlowski et al., "Prediction of Preterm Birth with and without Preeclampsia Using Mid-Pregnancy Immune and Growth-Related Molecular Factors and Maternal Characteristics," *Journal of Perinatology* 38, no. 8 (May 2018): 963–72.

37   called tocolytics: Information on medications to slow labor or delay birth found at "Treatments for Premature Labor," March of Dimes, accessed March 23, 2019, https://www.marchofdimes.org/complications/treatments-for-preterm-labor.aspx.

37   twenty-four hours to take effect: Clarissa Bonanno and Ronald J. Wapner, "Antenatal Corticosteroids in the Management of Preterm Birth: Are We Back Where We Started?" *Obstetrics and Gynecology Clinics of North America* 39, no. 1 (March 2012): 47–63, https://doi.org/10.1016/j.ogc.2011.12.006.

37   by gestational age: "Preterm Birth," World Health Organization, accessed March 23, 2019, https://www.who.int/news-room/fact-sheets/detail/preterm-birth.

38   similar categories: Clare L. Cutland et al., "Low Birth Weight: Case Definition & Guidelines for Data Collection, Analysis, and Presentation of Maternal Immunization Safety Data," *Vaccine* 35, no. 48 (December 2017): 6492–6500, https://doi.org/10.1016/j.vaccine.2017.01.049.

38   not always equivalents: John D. Lantos and Diane S. Lauderdale, *Preterm Babies, Fetal Patients, and Childbearing Choices* (Cambridge: MIT Press, 2015), 10.

2: TREATMENTS AND OUTCOMES

39   three crucial abilities: "Prematurity," Stanford University Children's Health, accessed March 23, 2019, https://www.stanfordchildrens.org/en/topic/default ?id=prematurity-90-P02401.

39   in the delivery room: Tom Lissauer et al., *Neonatology at a Glance* (West Sussex, UK: John Wiley & Sons, Ltd, 2016), 28–33.

40   both mortality and morbidity: R. E. Behrman and A. S. Butler, ed., Institute of Medicine (US) Committee on Understanding Premature Birth and Assuring Healthy Outcomes, *Preterm Birth: Causes, Consequences, and Prevention* (Washington, D.C.: National Academies Press, 2007), 10, https://www.ncbi.nlm.nih .gov/books/NBK11385/.

40   is idiosyncratic: "NIH Study Reveals Factors that Influence Premature Infant Survival, Disability," NIH, April 16, 2008, https://www.nih.gov/news-events /news-releases/nih-study-reveals-factors-influence-premature-infant-survival -disability.

40   doesn't make sense: "NICHD Neonatal Research Network (NRN): Extremely Preterm Birth Outcome Data," NIH NICHD, accessed March 25, 2019, https:// www1.nichd.nih.gov/epbo-calculator.

40   a blunt, population-level look: United States Department of Health and Human Services (US DHHS), Centers of Disease Control and Prevention (CDC), National Center for Health Statistics (NCHS), Division of Vital Statistics (DVS). "Linked Birth / Infant Death Records 2007–2016," as compiled from data provided by the fifty-seven vital statistics jurisdictions through the Vital Statistics Cooperative Program, on CDC WONDER Online Database, accessed March 25, 2019, http://wonder.cdc.gov/lbd-current.html.

41   Around 60 percent: Danielle Ehret et al., "Association of Antenatal Steroid Exposure with Survival Among Infants Receiving Postnatal Life Support at 22 to 25 Weeks' Gestation," *JAMA Network Open* 1, no. 6 (October 2018): figure 2, https://doi.org/10.1001/jamanetworkopen.2018.3235.

41   a motor disability: "Cerebral Palsy," Healthy Children, American Academy of Pediatrics, accessed March 26, 2019, https://www.healthychildren.org/English /health-issues/conditions/developmental-disabilities/pages/Cerebral-Palsy.aspx.

42   and an adjusted age: "Corrected Age for Preemies," Healthy Children, American Academy of Pediatrics, accessed March 26, 2019, https://www.healthychil dren.org/English/ages-stages/baby/preemie/Pages/Corrected-Age-For-Preemies .aspx.

43   according to some data: "NICHD Neonatal Research Network (NRN): Extremely Preterm Birth Outcome Data," National Institutes of Health's (NIH's) Eunice Kennedy Shriver National Institute of Child Health and Human Development (NICHD), accessed March 26, 2019, https://www.nichd.nih.gov/about/org/der /branches/ppb/programs/epbo/dataShow.

3: VIABILITY AND THE
ZONE OF PARENTAL DISCRETION

45   culture- and resource-specific: Neil Marlow, interview with the author, August 8, 2017.

45   rarely resuscitated: James Cummings, "Antenatal Counseling Regarding Resuscitation and Intensive Care Before 25 Weeks of Gestation," *Pediatrics* 136, no. 3 (September 2015), https://doi.org/10.1542/peds.2015-2336.

45   is a question with many answers: Neil Marlow, "The Elephant in the Deliv-

ery Room," *New England Journal of Medicine* 372, no. 19 (May 2015), 1856–57, https://doi.org/10.1056/NEJMe1502250.

46   very, very occasionally 21 weeks: Chrissy Cummings, interview with the author, January 30, 2018.

46   two of them survived: M. Norman et al., "Association Between Year of Birth and 1-Year Survival Among Extremely Preterm Infants in Sweden During 2004–2007 and 2014–2016," *Journal of the American Medical Association* 321, no. 12 (2019): 1188–99, https://doi.org/10.1001/jama.2019.2021.

46   Not everyone agrees: Lynn Gillam et al., "Decision-Making at the Borderline of Viability: Who Should Decide and on What Basis?" *Journal of Paediatrics and Child Health* 53, no. 2 (February 2017): 105–11, https://doi.org/10.1111/jpc.13423.

47   in poor countries do: "Preterm Birth," World Health Organization, accessed March 23, 2019, https://www.who.int/news-room/fact-sheets/detail/preterm-birth.

47   around 28 weeks and up: *Roe v. Wade*, 410, U.S., 160.

4: THE HISTORY OF INCUBATION

51   prematurely in 1802: Frank T. Marzials, *Life of Victor Hugo* (London: Walter Scott, 1888), 13.

51   Coney Island, 1903: David A. Sullivan, "Coney Island History: The Story of Thompson & Dundy's Luna Park," accessed March 27, 2019, http://www.heartof coneyisland.com/luna-park-coney-island.html.

52   emphasize their smallness: Claire Prentice, *Miracle at Coney Island* (Kindle Single, 2016), loc. 140.

52   Prussia and is now Poland: Dawn Raffel, *The Strange Case of Dr. Couney* (New York: Blue Rider Press, 2018), 23–25.

52   at the age of nineteen: Ibid., 33–35.

52   no formal treatment options: A. J. Liebling, "Patron of the Preemies," *New Yorker*, June 3, 1939, 22.

52   particular baby's progress: Ibid., 24.

52   has always inspired anxiety: Dr. Jeffrey P. Baker, interview with the author, September 14, 2017.

53   in the nineteenth century: Tina Cassidy, *Birth: The Surprising History of How We Are Born* (New York: Grove Press, 2006), 31.

53   may have known: Shannon Withycombe, *Lost: Miscarriage in Nineteenth-Century America* (New Brunswick: Rutgers University Press, 2018), 56.

53   textbook from 1683: Murdina M. Desmond, *Newborn Medicine and Society* (Austin: Eakin Press, 1998), introduction.

54   One report: Thomas Cone, *History of the Care and Feeding of the Premature Infant* (Philadelphia: Lippincott Williams & Wilkins, 1985), 9.

54   hang it over a lamp: Ibid., 9.

54   April 1815: this account reported in ibid., 18–19.

55   Sparta and Rome: Murdina M. Desmond, "Introduction," in *Newborn Medicine and Society* (Austin: Eakin Press, 1998), 19.

55   "tenderness taboo": Ibid.

56   for adults: William Meadow et al., "The Prediction and Cost of Futility in the NICU," *Acta Paediatrica* 101, no. 4 (April 2012): 397–402, https://doi.org/10.1111/j.1651-2227.2011.02555.x.

56   congenital disorders: Susan Schoon Eberly, "Fairies and the Folklore of Disability: Changelings, Hybrids and the Solitary Fairy," *Folklore* 99, no. 1 (1988): 58–77, https://doi.org/10.1080/015587X.1988.9716425.

56   were not human at all: Sandra Newman, "Infanticide," Aeon.co, November 27, 2017, https://aeon.co/essays/the-roots-of-infanticide-run-deep-and-begin-with -poverty.

56   modern concept of the premature baby: Jeffrey P. Baker, "Tarnier's Invention," in *The Machine in the Nursery* (Baltimore: Johns Hopkins University Press, 1996), 38–43.

56   fateful jaunt: PM Dunn, "Stéphane Tarnier (1828–1897), the Architect of Perinatology in France," *Archives of Disease in Childhood: Fetal and Neonatal Edition* 86, no. 2 (2002): 137–39, http://dx.doi.org/10.1136/fn.86.2.F137.

56   1878: Jeffrey P. Baker, *The Machine in the Nursery* (Baltimore: Johns Hopkins University Press, 1996), 26.

57   caused the fever: Ibid., 31–35.

57   plummeted: P. M. Dunn, "Stéphane Tarnier (1828–1897), the Architect of Perinatology in France," *Archives of Disease in Childhood: Fetal and Neonatal Edition* 86, no. 2 (2002): 137–139, http://dx.doi.org/10.1136/fn.86.2.F137.

57   jet-black three-piece suits: Ibid.

58   if he put a baby inside: Dawn Raffel, *The Strange Case of Dr. Couney* (New York: Blue Rider Press, 2018), 27.

58   wrapped in fleece: Jeffrey P. Baker, *The Machine in the Nursery* (Baltimore: Johns Hopkins University Press, 1996), 35.

58   sawdust-insulated walls: Mark Fraser, "Stéphane Tarnier and the Origin of Incubators for Premature Babies," presented at the 46th Annual General Meeting of the Scottish Society of the History of Medicine, October 24, 1994.

58   around 90 degrees: Jeffrey P. Baker, *The Machine in the Nursery* (Baltimore: Johns Hopkins University Press, 1996), 37.

58   every couple of hours: Dawn Raffel, *The Strange Case of Dr. Couney* (New York: Blue Rider Press, 2018), 29.

59   62 percent of them survived: Mark Fraser, "Stéphane Tarnier and the Origin of Incubators for Premature Babies," presented at the 46th Annual General Meeting of the Scottish Society of the History of Medicine, October 24, 1994.

59   In 1884: Jeffrey P. Baker, *The Machine in the Nursery* (Baltimore: Johns Hopkins University Press, 1996), 36.

59   presented their results this way: S. H. Sadler, *Infant Feeding by Artificial Means: A Scientific and Practical Treatise on the Dietetics of Infancy* (London: Routledge, 1909), 25.

60   a debatable idea: Jeffrey P. Baker, *The Machine in the Nursery* (Baltimore: Johns Hopkins University Press, 1996), 41.

60   were to blame!: Ibid., 45–46.

60   an article: "The Use of Incubators for Infants," *Lancet* 149, no. 3848 (May 1897): 1490–91.

61   and died: P. M. Dunn, "Stéphane Tarnier (1828–1897), the Architect of Perinatology in France," *Archives of Disease in Childhood: Fetal and Neonatal Edition* 86, no. 2 (2002): 137–39, http://dx.doi.org/10.1136/fn.86.2.F137.

61   from every angle: Jeffrey P. Baker, *The Machine in the Nursery* (Baltimore: Johns Hopkins University Press, 1996), 65.

61   were at odds: Ibid.

61   an illustration: Ibid., 96.

62   without human intervention: Ibid., 78.

62   a fancy restaurant oven: Thomas Cone, *History of the Care and Feeding of the Premature Infant* (Philadelphia: Lippincott Williams & Wilkins, 1985), 30.

62   fan for circulation: details on Lion incubator design: Ibid.

62   72 percent survival: Jeffrey P. Baker, *The Machine in the Nursery* (Baltimore: Johns Hopkins University Press, 1996), 88.

63   medical establishment: evolution of Lion's approach: Ibid., 87–92.

63   diamond jubilee: Dawn Raffel, *The Strange Case of Dr. Couney* (New York: Blue Rider Press, 2018), 39.

63   in tow: "Trans-Mississippi Exposition, Omaha, Nebraska, 1898," Neonatology on the Web, accessed April 2, 2019, http://www.neonatology.org/pinups/omaha.html.

63   undersized babies: A. J. Liebling, "Patron of the Preemies," *New Yorker*, June 3, 1939.

64   *Scientific American:* "Baby Incubators at the Pan American Exposition," *Scientific American*, August 3, 1901, 68.

65   incubator wards there: Jeffrey P. Baker, *The Machine in the Nursery* (Baltimore: Johns Hopkins University Press, 1996), 115.

66   around 1908: Information on DeLee: Ibid., 115–21.

66   one article: Joseph B. De Lee, M.D., "Infant Incubation, with the Presentation of a New Incubator and a Description of the System at the Chicago Lying-in Hospital," *Chicago Medical Recorder*, 1902, accessed April 2, 2019, http://www.neonatology.org/classics/delee/delee.html.

67   Cary Grant: Information on establishment of the clinic at Luna Park: Claire Prentice, *Miracle at Coney Island* (Kindle Single, 2016).

67   in today's dollars: Ibid., loc. 434.

68   for the show: Details on Couney's clinic and how he ran it from the books of Baker, Liebling, Prentice, and Raffel.

68   propaganda for preemies: A. J. Liebling, "Patron of the Preemies," *New Yorker*, June 3, 1939.

68   or shortly thereafter: Ibid.

68   said: Dawn Raffel, email to the author, December 16, 2018.

69   through their noses: Dawn Raffel, *The Strange Case of Dr. Couney* (New York: Blue Rider Press, 2018), 17.

69   how small it was: Claire Prentice, *Miracle at Coney Island* (Kindle Single, 2016), loc. 253.

69   Couney successfully treated: Many details from Allen's story: "Beth Allen," Couney Island History Project, oral history, June 28, 2007, accessed April 2, 2019, http://www.coneyislandhistory.org/oral-history-archive/beth-allen.

69   didn't want to: Clare Prentice, "How One Man Saved a Generation of Premature Babies," BBC News Magazine, May 23, 2016.

69   she told me: Beth Allen, email to the author, December 29, 2016.

69   she might have thought: Dawn Raffel, *The Strange Case of Dr. Couney* (New York: Blue Rider Press, 2018), 90.

70   "Better Babies": Steven Selden "Transforming Better Babies into Fitter Families: Archival Resources and the History of the American Eugenics Movement, 1908–1930," *Proceedings of the American Philosophical Society* 149, no. 2 (2005): 199–225, http://www.jstor.org/stable/4598925.

70   graded children: Dawn Raffel, *The Strange Case of Dr. Couney* (New York: Blue Rider Press, 2018), 132–33.

70   respectable endeavor: Basic ideas in this paragraph: Dr. Jeffrey P. Baker, interview with the author, September 14, 2017.

70   Atlantic City: Dawn Raffel, *The Strange Case of Dr. Couney* (New York: Blue Rider Press, 2018), 93.

70 his own incubator ward: Peter M. Dunn, "Julius Hess, MD (1876–1955), and the Premature Infant," *Archives of Disease in Childhood: Fetal and Neonatal Edition* 85, no. 2 (2001): F141–44, http://dx.doi.org/10.1136/fn.85.2.F141.

71 *New Yorker* profile: A. J. Liebling, "Patron of the Preemies," *New Yorker*, June 3, 1939.

71 any of those things beforehand: Claire Prentice, *Miracle at Coney Island* (Kindle Single, 2016).

72 That same year: Rebecca Rego Barry, "Coney Island's Incubator Babies," JSTOR Daily, August 15, 2018, https://daily.jstor.org/coney-islands-incubator-babies/.

72 health issue: Information on Silverman and birth certificate data: William A. Silverman, MD, interviewed by Lawrence M. Gartner, MD, June 10, 1997, American Academy of Pediatrics Oral History Project.

72 two-thirds of Americans: Aaron E. Carroll "The Real Reason the U.S. Has Employer-Sponsored Health Insurance," *New York Times*, September 5, 2017.

72 signed in 1965: "History," Centers for Medicare and Medicaid Services, accessed April 3, 2019, https://www.cms.gov/About-CMS/Agency-information/History/.

73 "started to pay": All information on neonatal insurance coverage from Louis Gluck, MD, interviewed by Lawrence M. Gartner, MD, February 21, 1997, American Academy of Pediatrics Oral History Project.

5: THE MODERN INCUBATOR, OR HOW TO BUILD A GIRAFFE

74 minute adjustments: All information on the Giraffe in this chapter from multiple in-person and phone interviews between the author and Karen Starr, Majid Akhavan, and Steve Falk.

6: THE INCUBATORS OF THE FUTURE

83 published a study: Emily A. Partridge et al., "An Extra-Uterine System to Physiologically Support the Extreme Premature Lamb," *Nature Communications* 8, no. 15112 (April 2017), https://doi.org/10.1038/ncomms15112.

83 a video: Meera Senthilingam, "'Plastic Bag Womb' Could Help Keep Premature Babies Alive," CNN, April 26, 2017, https://www.cnn.com/2017/04/26/health/artificial-womb-premature-babies-lambs/.

84 said that the main idea behind the biobag: All quotes and general insight from Kevin Dysart throughout this chapter from an in-person interview with the author at Children's Hospital of Philadelphia, June 2, 2017.

85 granted in 1955: Artificial Uterus, patent 2,723,660, E. M. Greenberg, November 15, 1955.

86 Flake said: "Recreating the Womb: New Hope for Premature Babies," Children's Hospital of Philadelphia, accessed April 3, 2019: https://www.chop.edu/video/recreating-womb-new-hope-premature-babies.

86 1987 described: Yoshinori Kuwabara et al., "Development of Extrauterine Fetal Incubation System Using Extracorporeal Membrane Oxygenator," *Artificial Organs* 11, no. 3 (June 1987): 224–27, https://doi.org/10.1111/j.1525-1594.1987.tb02663.x.

86 strike a reassuring tone: Masahiro Sakata et al., "A New Artificial Placenta with a Centrifugal Pump: Long-term Total Extrauterine Support of Goat Fetuses," *Journal of Thoracic and Cardiovascular Surgery* 115, no. 5 (May 1998): 1023–31, https://doi.org/10.1016/S0022-5223(98)70401-5.

86 normal development: Nobuya Unno et al., "Development of an Artificial Placenta: Survival of Isolated Goat Fetuses for Three Weeks with Umbilical Arteriovenous Extracorporeal Membrane Oxygenation," *Artificial Organs* 17, no. 12 (December 1993): 996–1003, https://doi.org/10.1111/j.1525-1594.1993.tb03181.x.

86 combined with parts: "Recreating the Womb: New Hope for Premature Babies," Children's Hospital of Philadelphia, accessed April 3, 2019: https://www.chop.edu/video/recreating-womb-new-hope-premature-babies.

87 previous attempts: discussion and citing of past attempts in Emily A. Partridge et al., "An Extra-Uterine System to Physiologically Support the Extreme Premature Lamb," *Nature Communications* 8, no. 15112 (April 2017), https://doi.org/10.1038/ncomms15112.

87 to pass through: Dysart interview.

87 new kind of umbilical cannula: Emily A. Partridge et al., "An Extra-Uterine System to Physiologically Support the Extreme Premature Lamb," *Nature Communications* 8, no. 15112 (April 2017), https://doi.org/10.1038/ncomms15112.

87 keep it sterile: Ibid.

87 three hundred gallons of it a day: "Recreating the Womb: New Hope for Premature Babies," Children's Hospital of Philadelphia, accessed April 3, 2019: https://www.chop.edu/video/recreating-womb-new-hope-premature-babies.

88 23 to 24 weeks: All details on the study design and results from Dysart interview and Postridge, "An Extra-Uterine System."

88 transfer them: Erika Engelhaupt, "Artificial Womb Could Offer New Hope for Premature Babies," *National Geographic*, April 25, 2017, https://news.nationalgeographic.com/2017/04/artificial-womb-lambs-premature-babies-health-science/.

89 piped in: Emily A. Partridge et al., "An Extra-Uterine System to Physiologically Support the Extreme Premature Lamb," *Nature Communications* 8, no. 15112 (April 2017), https://doi.org/10.1038/ncomms15112.

89 bioethicist said: Dr. John Lantos, in-person interview with the author, January 16, 2018.

89 echoed the sentiment: Steven McElroy, phone interview with the author, March 9, 2018.

89 research fascinating: Karen Starr, phone interview with the author on March 1, 2017, and in-person interview with the author, June 21, 2017.

90 said Dysart: Dysart interview.

90 he said: Flake responses to viability questions: "Recreating the Womb: New Hope for Premature Babies," Children's Hospital of Philadelphia, accessed April 3, 2019, https://www.chop.edu/video/recreating-womb-new-hope-premature-babies.

91 he said: Dysart interview.

91 article in *Maclean's*: Daniel Munro, "Are We Ready for the Brave New World of Artificial Wombs?" *Maclean's*, July 20, 2017, https://www.macleans.ca/society/health/are-we-ready-for-the-brave-new-world-of-artificial-wombs/.

91 wildly inaccurate: In fact, according to the NIH NICHD NRN data, an average-sized girl born at 25 weeks who has had the benefit of prenatal steroids has about an 80 percent chance of survival and 70 percent chance of survival without a severe impairment.

91 2018 story: George Winter, "The Brave New World of Wombless Gestation," *Irish Times,* October 17, 2018, https://www.irishtimes.com/life-and-style/health-family/the-brave-new-world-of-wombless-gestation-1.3657329.

92 she took issue: E. C. Romanis, "Artificial Womb Technology and the Frontiers of

Human Reproduction: Conceptual Differences and Potential Implications," *Journal of Medical Ethics* 44, no. 11 (2018): 751–55, http://dx.doi.org/10.1136/medethics-2018-104910.

93   The CHOP researchers referred to: Emily A. Partridge et al., "An Extra-Uterine System to Physiologically Support the Extreme Premature Lamb," *Nature Communications* 8, no. 15112 (April 2017), https://doi.org/10.1038/ncomms15112.

93   writes that: Jeffrey P. Baker, *The Machine in the Nursery* (Baltimore: Johns Hopkins University Press, 1996), 85.

94   the artificial womb: Katherine Don, "The High-Tech Future of the Uterus," *Atlantic*, January 5, 2015, https://www.theatlantic.com/health/archive/2015/01/the-high-tech-future-of-the-uterus/383232/.

94   *Popular Science* story: Gretchen Reynolds, "Artificial Wombs," *Popular Science*, August 1, 2005, https://www.popsci.com/scitech/article/2005-08/artificial-wombs.

95   a lot of media attention: Including, but not limited to the following: William Saletan, "The Mouse and the Rat," *Slate*, July 29, 2005, https://slate.com/technology/2005/07/the-mouse-and-the-rat.html; Soraya Chemaly, "What Do Artificial Wombs Mean for Women?" Rewire News, February 23, 2012, https://rewire.news/article/2012/02/23/what-do-artificial-wombs-mean-women/; Robin McKie, "Men Redundant? Now We Don't Need Women Either," *Guardian*, February 10, 2002, https://www.theguardian.com/world/2002/feb/10/medicalscience.research; Zoltan Istvan, "Artificial Wombs Are Coming, but the Controversy Is Already Here," Vice: Motherboard, August 4, 2014, https://motherboard.vice.com/en_us/article/8qx8kk/artificial-wombs-are-coming-and-the-controversys-already-here.

95   2015 article: Katherine Don, "The High-Tech Future of the Uterus," *Atlantic*, January 5, 2015, https://www.theatlantic.com/health/archive/2015/01/the-high-tech-future-of-the-uterus/383232/.

7: DR. MILDRED STAHLMAN
AND THE MINIATURE IRON LUNG

99   was born: Details of Martha Humphrey's (Lott's) birth from a phone interview between Lott and the author, November 29, 2017.

100   political tendencies: Details of Stahlman's early life and career throughout from Bill Snyder, "Intensive Caring: Stahlman Has Always Demanded Excellence from Herself, Others," *Vanderbilt University Medical Center Reporter*, https://www.mc.vanderbilt.edu/reporter/index.html?ID=3747.

100   hospital courtyard: Dr. Mildred Stahlman, in-person interview with the author, October 30, 2017.

100   until Martha's birth: Mildred Stahlman, MD, "Assisted Ventilation in Newborn Infants," in *Historical Review and Recent Advances in Neonatal and Perinatal Medicine,* ed. George F. Smith, MD, and Dharmapuri Vidyasagar, MD (Mead Johnson Nutritional Division, 1980), accessed at http://www.neonatology.org/classics/mj1980/ch15.html.

101   one baby too many: Factors that led to Stahlman being able to use the negative pressure ventilator for Martha Humphreys from Jeffrey P. Baker, "Technology in the Nursery," in *Formative Years*, ed. Alexandra Minna Stern and Howard Markel (Ann Arbor: University of Michigan Press, 2002), 78–79.

101   400 million: M. Ochs, "The Number of Alveoli in the Human Lung," *American Journal of Respiratory Critical Care Medicine* 169, no. 1 (January 2004): 120–24. http://doi.org/10.1164/rccm.200308-1107OC.

102 grayish-blue: S. Reuter, C. Moser, and M. Baack, "Respiratory Distress in the Newborn," *Pediatrics in Review* 35, no. 10 (October 2014): 417–29, http://doi.org/10.1542/pir.35-10-417.

102 may survive: background on the mechanics of RDS: "Respiratory Distress Syndrome," Stanford Children's Health, accessed April 3, 2019, https://www.stanford childrens.org/en/topic/default?id=respiratory-distress-syndrome-90-P02371.

102 premature baby boy: Charlotte Gray, *Reluctant Genius: Alexander Graham Bell and the Passion for Invention* (New York: Arcade Publishing, 2006), 222.

102 lived only three hours: Candice Millard, *Destiny of the Republic* (New York: Anchor Books, 2011), 258.

102 wrote in her diary: Charlotte Gray, *Reluctant Genius: Alexander Graham Bell and the Passion for Invention* (New York: Arcade Publishing, 2006), 222.

103 "Poor little one," she wrote: Details on Robert's birth and his parent's reactions: Ibid., 231–32.

103 baby's rib cage: Ibid., 223.

103 "described above." Quote and details on the vacuum jacket: Mildred Stahlman, MD, "Assisted Ventilation in Newborn Infants," in *Historical Review and Recent Advances in Neonatal and Perinatal Medicine,* ed. George F. Smith, MD, and Dharmapuri Vidyasagar, MD (Mead Johnson Nutritional Division, 1980), accessed at http://www.neonatology.org/classics/mj1980/ch15.html.

104 25,000 babies a year: Lawrence K. Altman, MD, "A Kennedy Baby's Life and Death," *New York Times,* July 29, 2013, https://www.nytimes.com/2013/07/30/health/a-kennedy-babys-life-and-death.html.

104 to every surface: Thomas Cone, *History of the Care and Feeding of the Premature Infant* (Philadelphia: Lippincott Williams & Wilkins, 1985), 117.

104 absence of surfactant: M. E. Avery and J. Mead, "Surface Properties in Relation to Atelectasis and Hyaline Membrane Disease," *AMA American Journal of Diseases of Childhood* 97 (May 1959): 517–23, http://doi.org/10.1001/archpedi.1959.02070010519001.

104 when oxygen was used: William A. Silverman, *Retrolental Fibroplasia: A Modern Parable* (New York: Grune & Stratton, 1980), 45.

104 injecting babies: Ibid., 43.

105 a new idea: Context around the evolution of oxygen treatment: Ibid., 43–90.

106 and Baltimore: Anecdote about Clifford and the cities most affected: Ibid., preface and 1–7.

106 10,000 children: Ibid., preface.

106 Stevie Wonder among them: Stevie Wonder, "Larry King Live," interview by Larry King, CNN, December 5, 2010.

106 fully understood: "Facts About Retinopathy of Prematurity (ROP)," National Eye Institute, accessed on April 5, 2019, https://nei.nih.gov/health/rop/rop.

107 likely to cause death: Details on these experimental treatments for RDS: Thomas Cone, *History of the Care and Feeding of the Premature Infant* (Philadelphia: Lippincott Williams & Wilkins, 1985), 125–30.

107 breathe for the babies: Very early ventilation strategies, publications: Ian Donald and Josephine Lord, "Augmented Respiration Studies in Atelectasis Neonatorum," *Lancet* 261, no. 6749 (January 1953): 9–14, https://doi.org/10.1016/S0140-6736(53)92511-2. Francis Benson et al., "Positive-Pressure Respirator Treatment of Severe Pulmonary Insufficiency in the Newborn Infant," *Acta Anaesthesiologica Scandinavica* 2, no. 1 (April 1958): 37–43, https://doi.org/10.1111/j.1399-6576.1958.tb05249.x.

107 wrote about it: Mildred Stahlman, MD, "Assisted Ventilation in Newborn Infants," in *Historical Review and Recent Advances in Neonatal and Perinatal Medicine*, ed. George F. Smith, MD, and Dharmapuri Vidyasagar, MD (Mead Johnson Nutritional Division, 1980), accessed at http://www.neonatology.org/classics /mj1980/ch15.html.

108 present data: Mildred Stahlman et al., "The Management of Respiratory Failure in the Idiopathic Respiratory Distress Syndrome of Prematurity," *Annals of the New York Academy of Sciences* 121, no. 3 (March 1965): 930–41, https://doi .org/10.1111/j.1749-6632.1965.tb14263.x.

109 eighty pediatricians: "Medical Scientist Training Program: Mildred Stahlman," Vanderbilt University Medical School, accessed April 5, 2019, https://medschool .vanderbilt.edu/mstp/stahlman-thomas-college/.

109 couldn't breathe: All quotes and context from Dr. Håkan Sundell from a phone interview with the author, November 27, 2017.

110 uncompromising reputation: Many details of Stahlman's personality, work, and place in the community from in-person interview between biographer Corey N. Reese and the author on October 30 and phone interviews on October 27, 2017, and March 22, 2019.

110 detail wrong: Context around the Vanderbilt NICU and Stahlman's activities and reputation from the author's in-person reporting trip, October 29–30, 2017.

8: DR. MARIA DELIVORIA-PAPADOPOULOS
AND THE RUGGED MACHINE

112 not practical: Jeffrey P. Baker, "Technology in the Nursery," in *Formative Years,* ed. Alexandra Minna Stern and Howard Markel (Ann Arbor: University of Michigan Press, 2002), 82.

113 Maria Delivoria: All quotes from and details about Delivoria-Papadopoulos in this chapter from an in-person interview between Delivoria-Papadopoulos and the author on June 2, 2017, and a phone interview on February 11, 2019.

114 had discovered: Tim Hilchey, "Dr. Irving Wexler, Pioneer in Blood Exchange, Dies at 86," *New York Times,* April 6, 1997.

117 the first survivor: Maria Delivoria-Papadopoulos and Paul R. Swyer, "Assisted Ventilation in Terminal Hyaline Membrane Disease," *Archives of Disease in Childhood* 39, no. 207 (October 1964): 481–84, http://dx.doi.org/10.1136/adc.39.207.481.

118 May 1963: Alistair G. S. Philip, "Forty Years of Mechanical Ventilation . . . Then and Now . . ." *NeoReviews* 4, no. 12 (December 2003): e335–39, https://doi.org /10.1542/neo.4-12-e335.

118 seven survivors: M. Delivoria-Papadopoulos, H. Levison, and P. Swyer, "Intermittent Positive Pressure Respiration as a Treatment in Severe Respiratory Distress Syndrome," *Archives of Disease in Childhood* 40, no. 213 (October 1965): 474–79, https://www.ncbi.nlm.nih.gov/pmc/articles/PMC2019449/.

118 early work: Dr. John Lantos, phone interview with the author, July 5, 2017.

120 partially collapsed: These details are from Mira's medical record, obtained from NYU Langone Medical Center.

9: JFK'S LOST BABY AND
THE ADVENT OF SURFACTANT

121 Knew what they were: Details of this day from the Associated Press, "It Started Out as Cape Outing," *Boston Globe,* August 9, 1963.

121 baptized immediately: Details of Patrick's appearance and the events directly after his birth from Steven Levingston, *The Kennedy Baby: The Loss that Transformed JFK* (New York: Diversion Books, 2013), Washington Post e-book.

121 a blue cast: Details of Patrick's condition and the chronology of care throughout this chapter from Michael S. Ryan, RRT-NPS, *Patrick Bouvier Kennedy: A Brief Life that Changed the History of Newborn Care* (Minneapolis: MCP Books, 2015).

122 could be done: Various details of the birth and JFK's arrival to Otis AFB: Douglas S. Crocket, "Baby Sped to Boston," *Boston Globe*, August 8, 1963.

122 on the news: Details about television, and also about Patrick's hair: Thurston Clarke, "A Death in the First Family," *Vanity Fair*, July 1, 2013, https://www.vanityfair.com/news/politics/2013/07/icebergs-jfk-jackie-death-patrick.

122 look at him now: Michael S. Ryan, RRT-NPS, *Patrick Bouvier Kennedy: A Brief Life that Changed the History of Newborn Care* (Minneapolis: MCP Books, 2015), 58–59.

123 at that time: Ibid., interview between Ryan and Drorbaugh, 96–107.

123 well-wishers outside: Ibid., 61–70.

124 seemed to stabilize: Details of Patrick's blood gases and treatment in ibid., 96–107.

124 phrase it: Douglas S. Crocket, "Baby Sped to Boston," *Boston Globe*, August 8, 1963.

124 "He's a Kennedy—He'll Make It": Mary McGrory, "He's a Kennedy—He'll Make It," *Boston Globe*, August 8, 1963.

124 an ode: Gloria Negri, "Whole World Taken by Littlest Kennedy," *Boston Globe*, August 8, 1963.

124 100 percent oxygen: *Patrick Bouvier Kennedy: A Brief Life that Changed the History of Newborn Care* (Minneapolis: MCP Books, 2015), 73.

125 Upper East Side: Rachel Zimmerman, "Tale of the Pediatrician Snatched to Treat the Kennedy Baby," WBUR, August 6, 2013, https://www.wbur.org/commonhealth/2013/08/06/tale-of-the-pediatrician-snatched-to-treat-kennedy-baby#more-33257.

125 say that someone called: Deliviora-Papadopoulos and Stahlman, in-person interviews with the author.

126 in his later interview with Ryan, said that: Michael S. Ryan, RRT-NPS, *Patrick Bouvier Kennedy: A Brief Life that Changed the History of Newborn Care* (Minneapolis: MCP Books, 2015), 114.

127 last time: details of the tank and Patrick's death: Ibid., 60–80.

127 so undone: J. Randy Taraborrelli, *Jackie, Janet & Lee* (New York: St. Martin's Press, 2018), 167.

128 was gone: Michael S. Ryan, RRT-NPS, *Patrick Bouvier Kennedy: A Brief Life that Changed the History of Newborn Care* (Minneapolis: MCP Books, 2015), 71–81.

128 "A Little Boy": *New York Times* Editorial Board, "A Little Boy," August 10, 1963.

129 Drorbaugh recalled: Michael S. Ryan, RRT-NPS, *Patrick Bouvier Kennedy: A Brief Life that Changed the History of Newborn Care* (Minneapolis: MCP Books, 2015), 106.

129 made for newborns: Patricia Bauer, "Forrest Morton Bird," *Encyclopaedia Britannica*, https://www.britannica.com/biography/Forrest-Morton-Bird.

129 synthetic surfactant: Marshall Klaus, MD, interviewed by Lawrence M. Gartner, MD, The American Academy of Pediatricians Oral History Project, January 7, 2000, https://www.aap.org/en-us/about-the-aap/Pediatric-History-Center/Documents/Klaus.pdf.

129 In 1967, he and other colleagues: J. Chu et al., "Neonatal Pulmonary Ischemia,"

Pediatrics 40, no. 4 (October 1967): 709–82, https://pediatrics.aappublications .org/content/40/4/709.

129   It didn't work: Henry L. Halliday, "The Fascinating Story of Surfactant," *Journal of Paediatrics and Child Health* 53, no. 4 (April 2017): 327–32, http://doi .org/10.1111/jpc.13500.

130   in amazement: Ibid.

130   difficult process: Eric Boodman, "How an Inconspicuous Slaughterhouse Keeps the World's Premature Babies Alive," Stat News, March 12, 2018, https://www .statnews.com/2018/03/12/cow-surfactant-premature-babies.

10: THE REVOLUTIONARY PRACTICE
OF LISTENING TO PREEMIES

133   Dr. Heidelise Als: All quotes and details on Als's work and life throughout this chapter from an in-person interview between Als and the author, November 1, 2018.

134   one summer afternoon: All Whitton quotes and details on the Willis-Knighton NICU from an in-person visit and interview between Whitton and the author on August 10, 2018.

135   developing brains: Some new research does suggest that private rooms might actually go too far in the other direction and be too silent and isolated, but this seems to only be a factor in the absence of family involvement.

135   short and long term: "NICU Private Rooms Improve Outcomes," Brown University Center for the Child at Risk, November 17, 2014, https://www.brown.edu /research/projects/children-at-risk/news/2014-11/nicu-private-rooms-improve -outcomes.

135   until 1987: Philip M. Boffey, "Infants' Sense of Pain Is Recognized, Finally," *New York Times*, November 24, 1987, https://www.nytimes.com/1987/11/24/science /infants-sense-of-pain-is-recognized-finally.html.

136   Sklamberg remembered: Felice Sklamberg, phone interview with the author on February 22, 2017.

137   under construction: Information on fetal brain development and chronology: "When Does the Fetus's Brain Begin to Work?" Zero to Three, accessed April 5, 2019, https://www.zerotothree.org/resources/1375-when-does-the-fetus-s-brain -begin-to-work.

137   center of the brain: Katherine Zhou, "Evolution of the Cerebral Cortex Makes Us Human," *Yale Scientific*, September 1, 2010, http://www.yalescientific.org /2010/09/evolution-of-the-cerebral-cortex-makes-us-human/.

138   stacking themselves up: Maureen Mulligan LaRossa and Sheena L. Carter, "Understanding How the Brain Develops," Emory University School of Medicine, http://www.pediatrics.emory.edu/divisions/neonatology/dpc/brain.html.

138   stack correctly: "Development of the Cerebral Cortex," Science Direct, https:// www.sciencedirect.com/topics/neuroscience/development-of-the-cerebral -cortex.

138   cognitive problems: Gregory Z. Tau and Bradley S. Peterson, "Normal Development of Brain Circuits," *Neuropsychopharmacology* 35, no. 1 (January 2010): 147–68, http://doi.org/10.1038/npp.2009.115.

141   a different etiology: Alex Griffel, "Preemies at High Risk of Autism Don't Show Typical Signs of Disorder in Early Infancy," Washington University in St. Louis, July 1, 2015, https://source.wustl.edu/2015/07/preemies-at-high-risk-of-autism -dont-show-typical-signs-of-disorder-in-early-infancy/.

148 He said it is a gentle place: James Helm, phone interview with the author, October 3, 2018.

150 Rogers says: Dr. Elizabeth Rogers, phone interview with the author, December 5, 2018.

151 occupational therapist: Felice Sklamberg, phone interview with the author, February 22, 2017.

11: FOLLOW-UP CARE

158 Follow-up care for survivors is not: John D. Lantos, "Cruel Calculus: Why Saving Premature Babies Is Better Business Than Helping Them Thrive," *Health Affairs* 29, no. 11 (November 2010): 2114–17, http://doi.org/10.1377/hlthaff.2009.0897.

159 much higher risks: J. L. Cheong et al., "Association Between Moderate and Late Preterm Birth and Neurodevelopment and Social-Emotional Development at Age 2 Years," *JAMA Pediatrics* 171, no. 4 (April 2017), https://doi.org/10.1001/jama pediatrics.2016.4805.

159 18 percent: March of Dimes Perinatal Data Center, "2018 Report Card: Louisiana," accessed April 5, 2019, https://www.marchofdimes.org/peristats/tools /reportcard.aspx?frmodrc=1&reg=22.

160 Louisiana Department of Health's: "Caddo Parish, Louisiana 2012–2014 Maternal and Child Health Profile," Louisiana Department of Health, accessed April 5, 2019, http://ldh.la.gov/assets/oph/Center-PHCH/Center-PH/maternal/Indicator Profiles/Region7Parishes_12-14/Caddo_2012-2014.pdf.

160 that of Malawi: "Malawi: Giving the Smallest Babies the Best Chance at Life," World Health Organization, accessed April 5, 2019, https://www.who.int/features /2015/malawi-infant-survival/en/.

160 Twenty-six percent: "Quick Facts: Caddo Parish, Louisiana," U.S. Census Bureau, accessed April 5, 2019, https://www.census.gov/quickfacts/fact/table/caddo parishlouisiana/IPE120217#IPE120217.

160 fifty babies: All information on Willis-Knighton NICU from Dr. Gerald Whitton, in-person interview with the author on August 10, 2018.

161 until they are four: Dr. Angela Montgomery, phone interview with the author, April 23, 2018.

161 she said: Context and quotes from Amy Sudduth throughout this chapter from phone interview with the author, October 25, 2018.

161 clinic's pediatric gym: All observations and quotes from inside the clinic throughout this chapter from an in-person reporting trip by the author on August 10, 2018.

172 summarized this: Kesha Evans, phone interview with the author, June 6, 2018.

12: WHAT SHOULD WE DO FOR 22-WEEK BABIES?

177 University of Iowa: All context, quotes, details, and information in this chapter obtained on the author's reporting trip to Iowa on February 25–27, 2018.

178 was flying: All quotes and context about the Hutchinson family from the author's visit to their home and interview with Chrissy and Jordan Hutchinson, February 25, 2018.

179 St. Luke's confirmed: In an email to the author, December 13, 2018.

180 told me later: Dr. Jonathan Klein, in-person interview with the author, February 26, 2018.

184 About 5,000: United States Department of Health and Human Services (US

DHHS), Centers of Disease Control and Prevention (CDC), National Center for Health Statistics (NCHS), Division of Vital Statistics (DVS). Linked Birth / Infant Death Records 2007–2016, as compiled from data provided by the fifty-seven vital statistics jurisdictions through the Vital Statistics Cooperative Program, on CDC WONDER Online Database, accessed April 8, 2019 at http://wonder.cdc.gov/lbd-current.html.

184 Her water broke: Lametria Burks, phone interview with the author, June 27, 2018.

185 Dr. Matthew Rysavy: Context around Rysavy's research and quotes from a phone interview between Rysavy and the author on August 9, 2017.

185 this puzzle: The data he was looking at was from the National Institute of Child Health and Human Development Neonatal Research Network, https://www.nichd.nih.gov/about/org/der/branches/ppb/programs/epbo.

186 separated out the numbers: Matthew A. Rysavy et al., "Between-Hospital Variation in Treatment and Outcomes in Extremely Preterm Infants," *New England Journal of Medicine* 372, no. 19 (May 2015): 1801–11, https://doi.org/10.1056/NEJMoa1410689.

186 a newer study: Danielle E. Y. Ehret et al., "Exposure with Survival Among Infants Receiving Postnatal Life Support at 22 to 25 Weeks' Gestation," *JAMA Network Open* 1, no. 6 (October 2018), https://doi.org/10.1001/jamanetworkopen.2018.3235.

187 industrialized nations: Úrsula Guillén et al., "Guidelines for the Management of Extremely Premature Deliveries: A Systematic Review," *Pediatrics* 136, no. 2 (August 2015): 343–50, https://doi.org/10.1542/peds.2015-0542.

187 in Japan: Yumi Kono et al., "Changes in Survival and Neurodevelopmental Outcomes of Infants Born at <25 Weeks' Gestation: A Retrospective Observational Study in Tertiary Centres in Japan," *Neonatology* 2, no. 1 (January 2018), https://doi.org/10.1136/bmjpo-2017-000211.

187 53 percent: Carl H. Backes et al., "Outcomes Following a Comprehensive Versus a Selective Approach for Infants Born at 22 Weeks of Gestation," *Journal of Perinatology* 39, no. 1 (October 2018): 39–47, https://doi.org/10.1038/s41372-018-0248-y.

188 10 percent: Dr. Tarah Colaizy, statistics provided to the author in an email, February 27, 2018.

188 shaping the policy: Details and quotes from Bell about his life, career, and clinical experience from phone interviews on January 2 and 12, 2018, and February 4, 2019, as well as in-person interviews in Iowa City on February 26 and 27, 2018.

192 noticed a generational divide: Dr. Brownsyne Tucker Edmonds, phone interviews with the author March 16, 2018, April 16, 2018, and April 30, 2018.

194 2017 study showed: Ravi M. Patel, "Survival of Infants Born at Periviable Gestational Ages," *Clinics in Perinatology* 44, no. 2 (June 2017): Figure 5, https://doi.org/10.1016/j.clp.2017.01.009.

195 "let die mistake": Dominic Wilkinson, *Death or Disability?* (Oxford: Oxford University Press, 2014), 246.

195 put it more strongly: Dr. John Dagle, in-person interview with the author, February 27, 2018.

196 also thinks: Dr. Jonathan Klein, in-person interview with the author, February 26, 2018.

196 does not recommend: American College of Obstetricians and Gynecologists and the Society for Maternal-Fetal Medicine, "Obstetric Care Consensus: Periviable Birth," *Obstetrics & Gynecology* 130, no. 4 (October 2017): e187–99, https://

www.acog.org/-/media/Obstetric-Care-Consensus-Series/occ006.pdf?dmc=1&ts
=20180129T0128313960.

196 according to a new study: Danielle E. Y. Ehret et al., "Exposure with Survival Among Infants Receiving Postnatal Life Support at 22 to 25 Weeks' Gestation," *JAMA Network Open* 1, no. 6 (October 2018), https://doi.org/10.1001/jama networkopen.2018.3235.

197 One study in Europe: Dominic Wilkinson, Eduard Verhagen, and Stefan Johansson, "Thresholds for Resuscitation of Extremely Preterm Infants in the UK, Sweden, and Netherlands," *Pediatrics* 142, no. 1 (September 2018): S574–84, https://doi.org/10.1542/peds.2018-0478I.

197 in the United States: F. McKenzie, B. K. Robinson, and B. Tucker Edmonds, "Do Maternal Characteristics Influence Maternal–Fetal Medicine Physicians' Willingness to Intervene When Managing Periviable Deliveries?" *Journal of Perinatology* 36 (March 2016): 522–28, https://doi.org/10.1038/jp.2016.15.

197 she said: Dr. Tarah Colaizy, in-person interview with the author on February 27, 2018.

199 forty-one 22-week babies: Provided to the author in an email from Edward Bell, February 5, 2019.

200 feel that Bell: Dr. Joseph Kaempf, phone interview with the author, January 19, 2018.

201 Dr. Mark Mercurio: Dr. Mark Mercurio, phone interview with the author, April 11, 2018.

13: KNOWING WHEN TO STOP

203 shoulder to shoulder: All information, context, and quotes in this chapter about Maria Fareri Children's Hospital and Dr. LaGamma from a series of in-person reporting sessions in the hospital by the author: May 31, June 12, June 14, June 16, June 23, and July 12, 2017.

206 first described: Raymond S. Duff and A. G. M. Campbell, "Moral and Ethical Dilemmas in the Special-Care Nursery," *New England Journal of Medicine* 289, no. 17 (October 1973): 890–94, https://doi.org/10.1056/NEJM197310252891705.

206 published an interview: Beverly Kelsey, "Which Infants Should Live? Who Should Decide?" *Hastings Center Report* (April 1975), https://doi.org/10.2307/3560810.

208 60 percent of deaths: Julie Weiner et al., "How Infants Die in the Neonatal Intensive Care Unit: Trends from 1999 Through 2008," *Archives of Pediatrics and Adolescent Medicine* 165, no. 7 (July 2011): 630–34, https://doi.org/10.1001/arch pediatrics.2011.102.

208 He remembers: Dr. Brian Carter, phone interviews with the author on March 7, 2018, and March 16, 2018.

210 generally feel: Hazel E. McHaffie, Andrew J. Lyon, and Robert Hume, "Deciding on Treatment Limitation for Neonates: The Parents' Perspective," *European Journal of Pediatrics* 160, no. 6 (May 2001): 339–44, https://doi.org/10.1007/PL00008444.

210 longed for a child: Laura and Graham Shullman, phone interviews with the author on July 19, 2017, and July 24, 2017.

213 a similar experience: Alexis Kropp-Kwon, phone interview with the author, July 23, 2017.

215 Aleshia Jones gave birth: Aleshia Jones, in-person interview with the author on August 14, 2018.

216 Parravicini disagrees: All quotes, stories, and context about Parravicini and details about her comfort care program from in-person interviews between Parravicini and the author on May 22, 2017, and February 5, 2018, as well as the author's attendance of a comfort care training run by Parravicini and her colleagues on October 17, 2017, at New York Presbyterian.

218 about 10 percent: "What Is Trisomy 18?" Trisomy 18 Foundation, accessed April 6, 2019, https://www.trisomy18.org/what-is-trisomy-18/.

219 In the article: Lyndsey R. Garbi, Shetal Shah, and Edmund F. LaGamma, "Delivery Room Hospice," *Acta Paediatrica* 105, no. 11 (November 2016): 1261–65, https://doi.org/10.1111/apa.13497.

221 constituted battery: Texas Supreme Court, "Miller v. HCA Inc.," *Wests South West Report* 118 (2003): 758–72, https://www.ncbi.nlm.nih.gov/pubmed/15765577.

221 emergency medical condition: John J. Paris, Michael D. Schreiber, and Frank Reardon, "The 'Emergent Circumstances' Exception to the Need for Consent: The Texas Supreme Court Ruling in *Miller v. HCA*," *Journal of Perinatology* 24 (May 2004): 337–42, https://doi.org/10.1038/sj.jp.7211105.

221 when they are gone: Jim Atkinson, "Wrongful Life?" *Texas Monthly*, December 2002, https://www.texasmonthly.com/articles/wrongful-life/.

223 Abraham told me about Aaron's life: Princy Abraham, in-person interview with the author, July 31, 2017.

227 consider qualifying: Brian S. Carter, "Pediatric Palliative Care in Infants and Neonates," *Children* 5, no. 2 (February 2018): 21, https://doi.org/10.3390/children 5020021.

14: CHOICE, DECISIONS,
AND THE MESSINESS OF REAL LIFE

227 and 3 days: All context and quotes around Annie Janvier's experience from a phone interview between Janvier and the author, March 23, 2018.

230 Research suggests: Brownsyne Tucker Edmonds et al., "A Pilot Study of Neonatologists' Decision-Making Roles in Delivery Room Resuscitation Counseling for Periviable Births," *AJOB Empirical Bioethics* 3 (July 2016): 175–82, http://dx.doi .org/10.1080/23294515.2015.1085460.

231 chance of survival: "NICHD Neonatal Research Network (NRN): Extremely Preterm Birth Outcome Data," National Institute of Child Health and Human Development, outcome data accessed April 6, 2019, https://www1.nichd.nih.gov /epbo-calculator/Pages/epbo_case.aspx.

232 "Nobody Likes Premies": A. Janvier, I. Leblanc, and K. Barrington, "Nobody Likes Premies: The Relative Value of Patients' Lives," *Journal of Perinatology* 28 (July 2008): 821–26, https://doi.org/10.1038/jp.2008.103.

233 It's possible: Ibid.

233 In another paper: A. Janvier and M. Mercurio, "Saving vs. Creating: Perceptions of Intensive Care at Different Ages and the Potential for Injustice," *Journal of Perinatology* 33 (April 2013): 333–35, https://doi.org/10.1038/jp.2012.134.

235 in another essay: Annie Janvier, "Pepperoni Pizza and Sex," *Current Problems in Pediatric and Adolescent Health Care* 41, no. 4 (April 2011): 106–8, https://doi .org/10.1016/j.cppeds.2010.11.002.

236 There is some evidence: Brownsyne Tucker Edmonds et al., "Racial and Ethnic Differences in Use of Intubation for Periviable Neonates," *Pediatrics* 127, no. 5 (May 2011): e1120–27, https://doi.org/10.1542/peds.2010-2608.

236 We walked into: All quotes, context, and details about Emma's case from an in-person interview with Dr. Jonathan Klein, February 27, 2018.
238 I asked Dr. Brian Carter: Dr. Brian Carter, phone interviews with the author on March 7, 2018, and March 16, 2018.

15: RACISM CAUSES PRETERM BIRTH

243 risk factor for preterm birth: Pathik D. Wadhwa et al., "Stress and Preterm Birth: Neuroendocrine, Immune/Inflammatory, and Vascular Mechanisms," *Maternal and Child Health Journal* 5, no. 2 (June 2001): 119–25, https://doi.org/10.1023/A:1011353216.

244 "allostatic load": David M. Olson et al., "Allostatic Load and Preterm Birth," *International Journal of Molecular Sciences* 16 no. 12 (December 2015): 29856–74, https://doi.org/10.3390/ijms161226209.

244 it is harmful: "Chronic Stress Puts Your Health at Risk," Mayo Clinic, accessed April 10, 2019, https://www.mayoclinic.org/healthy-lifestyle/stress-management/in-depth/stress/art-20046037.

244 hypertension: Feres José Mocayar Marón et al., "Hypertension Linked to Allostatic Load: From Psychosocial Stress to Inflammation and Mitochondrial Dysfunction," *Stress* 22, no. 2 (December 2018): 169–81, https://doi.org/10.1080/10253890.2018.1542683.

245 too much anxiety and fear: Gwen Latendresse, "The Interaction Between Chronic Stress and Pregnancy: Preterm Birth from a Biobehavioral Perspective," *Journal of Midwifery and Women's Health* 54, no. 1 (December 2010): 8–17, https://doi.org/10.1016/j.jmwh.2008.08.001.

245 studies have shown over and over: Jeanne L. Alhusen et al., "Racial Discrimination and Adverse Birth Outcomes: An Integrative Review," *Journal of Midwifery and Women's Health* 61, no. 6 (October 2016): 707–20, https://doi.org/10.1111/jmwh.12490.

245 a link between the experience: K. Bower et al., "Experiences of Racism and Preterm Birth: Findings from a Pregnancy Risk Assessment Monitoring System, 2004 through 2012," *Women's Health Issues* 28, no. 6 (November 2018): 495–501, https://doi.org/10.1016/j.whi.2018.06.002.

245 medical lists of risk factors: "What Are the Risk Factors for Preterm Labor and Birth?" National Institute of Child Health and Human Development, accessed April 10, 2019, https://www.nichd.nih.gov/health/topics/preterm/conditioninfo/who_risk.

245 she blamed herself: All quotes, context, and details about Crear-Perry's life and work from phone interviews with Crear-Perry on May 31, June 4, and June 13, 2018.

245 average preterm birth rate: Statistics in this paragraph from "2018 Premature Birth Report Card: United States," March of Dimes, accessed April 10, 2019, https://www.marchofdimes.org/materials/PrematureBirthReportCard-United%20States-2018.pdf; B. E. Hamilton et al., "Births: Provisional Data for 2018," Vital Statistics Rapid Release, no. 7 (2019), http://www.cdc.gov.

246 3 times more likely: "Very Preterm by Race/Ethnicity: United States, 2013–2015 Average," National Center for Health Statistics, final natality data, accessed April 10, 2019, www.marchofdimes.org/peristats.

246 children are more than twice: "Infant Mortality," CDC: Division of Reproductive Health, National Center for Chronic Disease Prevention and Health Promotion,

accessed April 10, 2019, https://www.cdc.gov/reproductivehealth/maternalinfant
health/infantmortality.htm.

246 control for socioeconomic status: Heather H. Burris and James W. Collins, "Commentary: Race and Preterm Birth—The Case for Epigenetic Inquiry," *Ethnicity & Disease* 20, no. 3 (June 2010): 296–99, https://www.ncbi.nlm.nih.gov/pubmed/20828105.

246 New York City: "Preterm by race/ethnicity: New York City, 2013–2015 Average," National Center for Health Statistics, final natality data, accessed April 10, 2019, www.marchofdimes.org/peristats.

247 nearly as much: Comparisons of risk in this paragraph: Crear-Perry, phone interview with the author on June 13, 2018.

247 behavioral factors: Heather H. Burris and James W. Collins, "Commentary: Race and Preterm Birth—The Case for Epigenetic Inquiry," *Ethnicity & Disease* 20, no. 3 (June 2010): 296–99, https://www.ncbi.nlm.nih.gov/pubmed/20828105. Additionally, the American Lung Association says that, for instance, more white women than Black women smoke cigarettes: https://www.lung.org/stop-smoking/smoking-facts/tobacco-use-racial-and-ethnic.html.

247 not a meaningful genetic category: Kacey Y. Eichelberger, Julianna G. Alson, and Kemi M. Doll, "Should Race Be Used as a Variable in Research on Preterm Birth?" *AMA Journal of Ethics* 20, no. 3 (March 2018): 296–302, https://doi.org/10.1001/journalofethics.2018.20.3.sect1-1803.

248 and vice versa: Siddhartha Mukherjee, *The Gene* (New York: Scribner, 2016), 341–42.

248 general health profile: Megan Gannon, "Race Is a Social Construct, Scientists Argue," *Scientific American*, February 5, 2016, https://www.scientificamerican.com/article/race-is-a-social-construct-scientists-argue.

248 two copies of a gene: Jefferson M. Fish, "Sickle Cell Anemia Isn't Evidence for the Existence of Races," *Psychology Today*, March 19, 2013, https://www.psychologytoday.com/us/blog/looking-in-the-cultural-mirror/201303/sickle-cell-anemia-isn-t-evidence-the-existence-races.

248 genetic contribution: Ge Zhang et al., "Genetic Associations with Gestational Duration and Spontaneous Preterm Birth," *New England Journal of Medicine* 377, (September 2017): 1156–67, https://doi.org/10.1056/NEJMoa1612665.

248 demonstrated the same thing: Richard J. David and James W. Collins Jr., "Differing Birth Weight Among Infants of U.S.-Born Blacks, African-Born Blacks, and U.S.-Born Whites," *New England Journal of Medicine* 337 (October 1997): 1209–14, https://doi.org/10.1056/NEJM199710233371706; Richard J. David and James W. Collins Jr., "Disparities in Infant Mortality: What's Genetics Got to Do with It?" *American Journal of Public Health* 97, no. 7 (July 2007): 1191–97, https://doi.org/10.2105/AJPH.2005.068387. One of the more recent of this kind of study was by Dr. Irina Buhimschi et al. in the *Journal of Obstetrics and Gynecology* in 2018. The study compiled vital records from Ohio over the years 2000 to 2015 to look at the prevalence of preterm birth among women in four groups: U.S.-born Black women, Africa-born Black women (excluding Arabic-speaking North African women), U.S.-born white women, and Somalia-born women, of which there is a large population in Ohio. In Ohio during this fifteen-year period, U.S.-born Black women gave birth early 13.3 percent of the time; Africa-born Black women, 8.4 percent of the time; U.S.-born white women, 7.9 percent of the time; Somalia-born women, 5.9 percent of the time.

249 said Dr. Arthur James: Dr. Arthur James, phone interview with the author, May 30, 2018.

250 the term "weathering": A. Geronimus, "The Weathering Hypothesis and the Health of African-American Women and Infants: Evidence and Speculations," *Ethnicity & Disease* 2 no. 3 (Summer 1992): 207–21, https://www.ncbi.nlm.nih.gov /pubmed/1467758.

250 real and measurable: Jo C. Phelan and Bruce G. Link, "Is Racism a Fundamental Cause of Inequalities in Health?" *Annual Review of Sociology* 41 (August 2015): 311–30, https://doi.org/10.1146/annurev-soc-073014-112305.

250 discrimination and premature birth: James W. Collins Jr. et al., "Very Low Birthweight in African American Infants: The Role of Maternal Exposure to Interpersonal Racial Discrimination," *American Journal of Public Health* 94, no. 12 (December 2004): 2132–38, https://www.ncbi.nlm.nih.gov/pmc/articles/PMC 1448603/; Tyan Parker Dominguez et al., "Racial Differences in Birth Outcomes: The Role of General, Pregnancy, and Racism Stress," *Journal of Health Psychology* 27, no. 2 (March 2008): 194–203, https://doi.org/10.1037/0278-6133.27.2.194; Paula Braveman et al., "Worry About Racial Discrimination: A Missing Piece of the Puzzle of Black-White Disparities in Preterm Birth?," *PLoS One* 12, no. 10 (October 2017): https://doi.org/10.1371/journal.pone.0186151.

250 stress hormone: R. Jeanne Ruiz et al., "CRH as a Predictor of Preterm Birth in Minority Women," *Biological Research for Nursing* 18, no. 3 (May 2016): 316–21, https://doi.org/10.1177/1099800415611248.

250 tightly associated with each other: James Collins, phone interview with the author, July 2, 2018.

250 structural racism: Brittany D. Chambers et al., "Using Index of Concentration at the Extremes as Indicators of Structural Racism to Evaluate the Association with Preterm Birth and Infant Mortality—California, 2011–2012," *Journal of Urban Health* 96, no. 2 (April 2019): 159–70, https://doi.org/10.1007/s11524-018 -0272-4.

251 recent study: Brittany D. Chambers et al., "Pregnant Black Women's Experiences of Racial Discrimination," *American Journal of Obstetrics & Gynecology* 218, no. 1 (January 2018): S569, https://doi.org/10.1016/j.ajog.2017.11.448.

251 The research tracks self-reported: Brittany D. Chambers, information on this study from a phone interview with the author, July 25, 2018.

251 her own life: All quotes, context, and details throughout this chapter about Crear-Perry's life and work from phone interviews with Crear-Perry on May 31, June 4, and June 13, 2018.

254 discriminatory health care: M. McLemore et al., "Health Care Experiences of Pregnant, Birthing and Postnatal Women of Color at Risk for Preterm Birth," *Social Science & Medicine* 201 (March 2018): 127–35, https://doi.org/10.1016/j .socscimed.2018.02.013.

255 women's symptoms: Laura Kiesel, "Women and Pain: Disparities in Experience and Treatment," Harvard Health Publishing, October 9, 2017, https://www .health.harvard.edu/blog/women-and-pain-disparities-in-experience-and-treat ment-2017100912562.

255 talked about the birth: Hope Williams, in-person interview with the author, November 19, 2017.

257 well documented: Elizabeth Brondolo et al., "Racism and Hypertension: A Review of the Empirical Evidence and Implications for Clinical Practice," *American Journal of Hypertension* 24, no. 5 (May 2011): 518–29, https://doi.org/10.1038 /ajh.2011.9.

258 on the cover: Linda Villarosa, "Why America's Black Mothers and Babies Are in

a Life-or-Death Crisis," *New York Times Magazine*, April 11, 2018, https://www
.nytimes.com/2018/04/11/magazine/black-mothers-babies-death-maternal
-mortality.html.

258   for weeks: Priska Neely, "Special Series: Black Infant Mortality," KPCC, https://
www.scpr.org/topics/special-series-black-infant-mortality.

258   he can feel it: James Collins, phone interview with the author, July 2, 2018.

16: WHAT PREMATURITY MEANS IN MISSISSIPPI

260   births are early: March of Dimes Perinatal Data Center, "2018 Premature Birth
Report Card: Mississippi," accessed April 10, 2019, https://www.marchofdimes
.org/peristats/tools/reportcard.aspx?frmodrc=1&reg=28.

260   not in denial about this: All details and quotes throughout this chapter about
the life and work of Charlene Collier as well as data from Mississippi from an in-
person interview between Collier and the author, August 14, 2018.

260   highest poverty rate: "Overall Poverty: 2018," Talk Poverty, accessed April 10,
2019: https://talkpoverty.org/indicator/listing/poverty/2018.

261   "maternity care deserts": Symphonie Privett, "Dangerous Deliveries: Why Are
Black Women Dying During Childbirth?" WLBT, October 31, 2018, http://
www.wlbt.com/2018/11/01/dangerous-deliveries-why-are-black-women-dying
-during-childbirth/.

261   to pregnant women: A recent *New York Times* investigation described the second-
-trimester miscarriages and premature births experienced by multiple women
doing heavy lifting in a Verizon contractor warehouse. They had all submitted
a doctor's note requesting lighter duties and had all been turned down. Jessica
Silver-Greenberg and Natalie Kitroeff, "Miscarrying at Work: The Physical Toll of
Pregnancy Discrimination," *New York Times*, October 21, 2018, https://www.ny
times.com/interactive/2018/10/21/business/pregnancy-discrimination-mis
carriages.html.

263   quadrupled since 1999: "The Number of Women with Opioid Use Disorder at
Labor and Delivery Quadrupled from 1999–2014," CDC Newsroom, accessed
April 10, 2019, https://www.cdc.gov/media/releases/2018/p0809-women-opiod
-use.html.

263   4.8 percent: Jennifer McClellan, "Number of Pregnant Women Addicted to Opioids Soared over 15 years, CDC Says," *USA Today,* August 10, 2018, https://www
.usatoday.com/story/life/allthemoms/2018/08/10/opioid-crisis-rate-pregnant
-women-addicted-has-soared-cdc-says/956717002/.

265   makes financial (and ethical) sense: Adam Searing and Donna Cohen Ross,
"Medicaid Expansion Fills Gaps in Maternal Health Coverage Leading to Healthier Mothers and Babies," Georgetown University Health Policy Institute: Center for Children and Families, May 2019, https://ccf.georgetown.edu/wp-content
/uploads/2019/05/Maternal-Health_FINAL-1.pdf

265   can't afford to stay open: Aallyah Wright, "Rural Hospitals at Risk, Pose Real
Challenges for Mississippi Leaders," *Mississippi Today*, October 26, 2018, https://
mississippitoday.org/2018/10/26/rural-hospitals-at-risk-pose-real-challenges
-to-mississippi-leaders-and-communities/.

265   expanded Medicaid: Shawn Radcliffe "Obamacare Helped Keep Rural Hospitals
Open," Healthline, January 16, 2018, https://www.healthline.com/health-news
/obamacare-helped-keep-rural-hospitals-open.

266   300,000 Mississippians: Aallyah Wright, "Rural Hospitals at Risk, Pose Real
Challenges for Mississippi Leaders," *Mississippi Today*, October 26, 2018, https://

mississippitoday.org/2018/10/26/rural-hospitals-at-risk-pose-real-challenges
-to-mississippi-leaders-and-communities/.

266 family of four: Mississippi Division of Medicaid: Income Limits for Medicaid and
CHIP Programs, 2019, https://medicaid.ms.gov/medicaid-coverage/who-qualifies
-for-coverage/income-limits-for-medicaid-and-chip-programs/.

266 every way they can: Arielle Dreher "Planned Parenthood Braces for Battle," *Jackson Free Press*, July 25, 2018: http://www.jacksonfreepress.com/news/2018/jul
/25/planned-parenthood-braces-battle/.

266 Reproductive Health: T. Thompson and J. Seymour, "Evaluating Priorities: Measuring Women's and Children's Health and Wellbeing Against Abortion Restrictions in the States," Ibis Reproductive Health, June 2017, https://ibis
reproductivehealth.org/sites/default/files/files/publications/Evaluating%20Priori
ties%20August%202017.pdf.

267 18 percent in Mississippi: Susan L. Hayes et al., "What's at Stake: States' Progress
on Health Coverage and Access to Care, 2013–2016," Commonwealth Fund, December 14, 2017, https://www.commonwealthfund.org/publications/issue-briefs
/2017/dec/whats-stake-states-progress-health-coverage-and-access-care-2013.

267 the lowest rates: Associated Press, "Survey Finds 3 Percent of Vermonters Lack
Health Insurance," *U.S. News & World Report*, January 10, 2019, https://www
.usnews.com/news/best-states/vermont/articles/2019-01-10/survey-finds-3-per
cent-of-vermonters-lack-health-insurance.

267 maternity care: O. M. Campbell, "What Works? A Peek into Vermont's Model of
Maternity Care," *Rural Health Quarterly*, September 6, 2018, http://ruralhealth
quarterly.com/home/2018/09/06/what-works-a-peek-into-vermonts-model-of
-maternity-care/.

267 restrictions on abortion: "State Facts About Abortion: Vermont," Guttmacher
Institute, accessed April 10, 2019, https://www.guttmacher.org/fact-sheet/state
-facts-about-abortion-vermont.

267 preterm birth rate: "Percentage of Births Born Preterm by State," CDC, accessed
April 10, 2019, https://www.cdc.gov/nchs/data/nvsr/nvsr67/nvsr67_08-508.pdf.

267 infant mortality rates: "Infant Mortality Rates by State," CDC, accessed April 10,
2019, https://www.cdc.gov/nchs/pressroom/sosmap/infant_mortality_rates/infant
_mortality.htm.

267 Thompson says of Collier: All quotes and details in this chapter on Thompson's
work from an in-person interview between Thompson and the author, August 13,
2018.

17: GROUP PRENATAL CARE
AND THE POWER OF COMMUNITY

271 4 percent: Internal statistics for BronxCare provided by Leah Halliburton in a
phone call on Wednesday, March 20, 2019.

272 a similar program: "March of Dimes, UnitedHealth Group Launch Group Prenatal Care Program to Help Improve Health Outcomes for Mothers and Babies,
and to Reduce Health Care Costs," March of Dimes, June 23, 2016, https://www
.marchofdimes.org/news/march-of-dimes-unitedhealth-group-launch-group-pre
natal-care-program-to-help-improve-health-outcomes-for-mothers-and-babies
-and-reduce-health-care-costs.aspx.

272 providers who use it: The information about UnitedHealthcare providing enhanced reimbursement was part of a discussion at a March of Dimes conference
in Washington, D.C., on May 21, 2018. United declined to comment further.

272 12 percent to 3.5 percent: Amy Crockett et al., "Investing in CenteringPregnancy Group Prenatal Care Reduces Newborn Hospitalization Costs," *Women's Health Issues* 27, no. 1 (January 2017): 60–66, https://doi.org/10.1016/j.whi.2016.09.009.

272 68 percent: Shayna D. Cunningham, et al., "Group Prenatal Care Reduces Risk of Preterm Birth and Low Birth Weight: A Matched Cohort Study," *Journal of Women's Health* 28, no. 1 (January 2019), https://doi.org/10.1089/jwh.2017.6817.

272 unpublished data: Data provided in an email from Tanya Monroe of CHI, May 17, 2018.

272 now offers: Wesley P. Elliott, "Centering Prenatal Care Around You," U.S. Army, November 18, 2016, https://www.army.mil/article/178532/centering_prenatal _care_around_you.

273 has so far been shown: Amy H. Picklesimer et al., "The Effect of CenteringPregnancy Group Prenatal Care on Preterm Birth in a Low-Income Population," *American Journal of Obstetrics and Gynecology* 206, no. 5 (May 2012): 415.e1–415.e7, https://doi.org/10.1016/j.ajog.2012.01.040.

273 thinks that: Monica McLemore, phone interview with the author, July 6, 2018.

273 series of CenteringPregnancy sessions: All details, quotes, and stories from the Centering group in this chapter come from in-person reporting done at the Centering sessions at Bronx-Lebanon on April 18, May 2, May 16, and August 15, 2018.

278 she has seen it herself: Additional context and quotes from Leah Halliburton, phone interview with the author on April 23, 2018.

280 one early birth: All information on Baby Steps from a phone interview with Micah Sandager, RN and program director, on June 5, 2018.

285 she thinks: Crear-Perry, phone interview with the author, June 4, 2018.

285 Collier thinks this is an important step: Collier interview with the author, August 14, 2018.

18: THE HIDDEN TRAUMA OF PREMATURITY

289 A recent study: J. L. Cheong et al., "Association Between Moderate and Late Preterm Birth and Neurodevelopment and Social-Emotional Development at Age 2 Years," *JAMA Pediatrics* 171, no. 4 (April 2017), https://doi.org/10.1001/jama pediatrics.2016.4805.

290 was staying overnight: Liz Rodriguez Sowek, phone interview with the author, March 8, 2019.

292 at higher risk: S. N. Vigod, "Prevalence and Risk Factors for Postpartum Depression Among Women with Preterm and Low-Birth-Weight Infants: A Systematic Review," *BJOG* 117, no. 5 (April 2010): 540–50, https://doi.org/10.1111/j.1471 -0528.2009.02493.x.

292 for PTSD: R. J. Shaw, "The Relationship Between Acute Stress Disorder and Posttraumatic Stress Disorder in the Neonatal Intensive Care Unit," *Psychosomatics* 50, no. 2, (March 2009): 131–37, https://doi.org/10.1176/appi.psy.50.2.131.

297 had a similar experience: Kristen Mascia, phone interview, March 1, 2019.

19: GROWN PREEMIES SPEAK FOR THEMSELVES

297 noticed that parents of very premature: Quotes and details throughout the chapter from Saroj Saigal, phone interview with the author, December 3, 2018, and from Saigal's book, *Preemie Voices* (Victoria, Canada: FriesenPress, 2014).

Index

abortion, 36, 197–99, 266–67
 decision-making in, 32
 and maternal and infant
  mortality rates, 266
 religion and, 197
 Roe v. Wade and, 27, 47, 197, 199
 viability and, 27, 46–47, 90,
  197–99
Abraham, Aaron, 222–26, 237
Abraham, Princy, 222–26, 237
adrenaline, 244
Affordable Care Act, 268
Africa, 248, 249
African Americans:
 health care and, 254–55
 hospice care and, 236
 life expectancy and, 250
 maternal and infant mortality in,
  34, 258
 in Mississippi, 261, 269
 prematurity and, 8, 34, 243–59,
  262, 263
 see also racism
age:
 adjusted, 42, 154, 295
 see also gestational age
Agia Sofia Children's Hospital, 113
Akhavan, Majid, 77–78, 79
Allen, Beth, 69
Als, Heidelise, 133–36, 138,
  141–52, 153, 165, 168
alveoli, 101, 102
American Academy of Pediatrics,
  135

American College of Obstetricians
  and Gynecologists (ACOG),
  196
amniotic sac, 85
amniotic fluid, 87, 101
anesthesia, 135
anti-Semitism, 64–65
artificial womb (biobag), 83–96
Asian women, 246
asphyxia, 203
aspirin, 37, 263, 264, 268, 270
asthma, 33, 158
Atlantic, 95
Auchincloss, Janet, 125, 127
autism, 140–41
Avery, Mary Ellen, 104

Baby Steps, 280
Baker, Jeffrey P., 58, 60, 65, 93
Bangladesh, 160
Barrington, Keith, 228, 232
Bell, Alexander Graham, 102–4,
  120
Bell, Edward, 188–95, 199–202
Bell, Mabel, 102–4
Bernhard, William F., 126–28
Better Babies craze, 70
biobag (artificial womb), 83–96
Bird ventilators, 116, 129
birth:
 early elective, 264
 medications to delay, 37
 "natural," emphasis on, 292
 stillbirth, 34

birth weight, 38, 40, 59, 72
blood transfusions, 114–15
bonding, 17, 91–92, 134
Booker, Cory, 258
Boston Children's Hospital,
    121–28, 134
Boston Globe, 124
bradycardia, 19, 23–24, 25, 209
brain, 17, 19, 85, 135–41, 234
    damage to, 203, 204, 234
    development of, 133, 136–37,
        150, 152
    neurons in, 138
    sensory stimulation and, 135
Brazelton, T. Berry, 147
breastfeeding, 155–56
breathing, 39, 102
    see also respiratory distress
    syndrome
bronchopulmonary dysplasia
    (BPD), 223–26, 236
Bronx-Lebanon Hospital
    (BronxCare), 271, 273, 278
Bryn, Julia, 163–65, 170–71
Budin, Pierre, 61, 62, 71
Buffalo Medical Journal, 65
Burks, Carli, 185
Burks, Lametria, 184–85

Cabbage Patch dolls, 296
Caddo Parish, La., 159–60
caffeine, 15, 16
Campbell, A. G. M., 206
Canada, 199
cannulas, nasal, 39, 223
Caribbean, 248
Carter, Brian, 208–10, 238–39
CDC, 40
CenteringPregnancy, 272–85
cerebral palsy (CP), 41–42, 43,
    136, 155, 157, 193, 221, 301,
    304, 305

cervix, 36
Chambers, Brittany D., 251
changeling idea, 56
chest compressions, 180, 208–10
Children's Center Rehabilitation
    Hospital, 238–39
Children's Hospital of Philadelphia
    (CHOP), 83, 85, 86–89, 91–93
Children's Mercy Kansas City, 208
China, 78
choices about care, *see* decisions
    and choices about care
chromosomal abnormalities, 3, 32,
    218
Chu, Jacqueline, 129
Clifford, Stewart, 105
Colaizy, Tarah, 197–98
Collier, Charlene, 260–65, 267,
    270, 285–86
Collins, James, 248–49, 250, 258
Columbia-Presbyterian Babies
    Hospital, 72
Columbia University Medical
    Center/New York-Presbyterian
    Morgan Stanley Children's
    Hospital, 217
comfort care, 41, 46, 179, 183, 192,
    200–201, 208, 209, 218, 222,
    225
communication, mother-baby,
    141–42, 145, 147
community, power of, 271–86
Coney Island, 51–52, 67–70, 71–72
contraception, 36, 266
Cornell Hospital, 72
Cornell University, 94–95
corticosteroids:
    stress and, 244
    treatment with, *see* steroid
    treatment
Couney, Annabelle Maye, 67
Couney, Martin, 52, 63–73, 93

CPAP (continuous positive airway pressure), 17, 25, 39, 223–25
CPR, 23
crawling, 156–57
Crear-Perry, Joia, 245, 247, 251–54, 259, 285
C-section, 35, 53, 197, 255, 264

Dagle, John, 195, 235
David, Richard, 248
death, 34, 39–40, 80, 233
    allowing baby to die, 41, 45, 203–26
    comfort care until, 41, 46, 179, 183, 192, 200–201, 208, 209, 218
    and decisions about continuing care, see decisions and choices about care
    by gestational age at birth, 40–41
    infanticide, 56
    maternal and infant mortality rates, 34, 258, 266, 267
    medications and, 220
    premature birth as cause of, 32–33
    stillbirth, 34
    trial of therapy and, 194–95, 204, 206
    withdrawing life support and, 194, 195, 203–26
    withholding resuscitation and, 189
decisions and choices about care, 47, 189–92, 198–200, 203–26, 227–39
    ethical questions in, 198–99, 233, 235–36
    for premature babies vs. other patients, 232–33
    quality of life and, 235, 239
DeLee, Joseph B., 65–67, 73

Delivoria-Papadopoulos, Maria, 112–20, 122, 123, 125, 129, 142, 143, 191
depression, 292
development, 153–54
    of brain, 133, 136–37, 150, 152
    crawling and, 156–57
    eyelids and, 190
    developmental care, 134, 143, 151, 154, 158
    follow-up, 153–73
disabilities, 41–44, 91, 155, 158, 186, 193, 207, 221, 231–32, 234, 289–90, 297–99, 301, 302, 305
DNA test, 3
Don, Katherine, 95
Down syndrome, 3, 32
Drexel University, 113, 119
Drorbaugh, James E., 121–27, 129
Duff, Raymond, 206–8
Dysart, Kevin, 84, 90, 91

Early Intervention System, 33, 42, 156, 157, 290
    Developmental Profile, 163
Early Steps, 162, 164
ECMO (extracorporeal membrane oxygenation) machine, 86–87, 208
education levels, 246, 247
egg warmers, 57–58
Einstein, Albert, 53
endotracheal tubes, 120, 183, 191
ethical questions in healthcare decisions, 198–99, 233, 235–36
eugenics, 64–65, 70
Eunice Kennedy Shriver National Institute of Child Health and Human Development, 43
Evans, Kesha, 172–73
Exposition Universelle, 63

Falk, Steve, 79–82
FDA, 96
feeding tubes, 39, 59, 60, 182–83
fetus, 92
    biobag and, 83–96
    personhood of, 46–47
Flake, Alan, 86, 89, 90–91
follow-up care, 153–73
Franco-Prussian War, 60

Gartner, Lawrence, 73
gavage, 59, 60
genetics:
    preterm birth and, 248–49
    race and, 247–48, 254
Geronimus, Arline, 250
gestational age, 35, 37–38, 40–41,
        53, 59, 72
    measuring, 195
    22-week babies, 18, 40, 45–46,
        88, 90, 91, 177–202
    23-week babies, 18, 40, 46, 88,
        89–91, 177, 184–88, 195,
        199–201, 206
    24-week babies, 40–41, 45–46,
        59, 80, 88, 137, 177, 232
Gillibrand, Kirsten, 258
Giraffe OmniBed Carestations,
        74–82
Gluck, Louis, 73
goats, 85, 86
Grant, Cary, 67
Gray, Charlotte, 103
Greater Baltimore Medical
        Center, 76
Ground Beneath Her Feet, The
        (Rushdie), 306
group prenatal care, 271–86

Haiti, 160
Halliburton, Leah, 274–85
Harris, Kamala, 258

Harvard University, 123
Hastings Center, 206
healthcare decisions, ethical
        questions in, 198–99, 233,
        235–36
health insurance, 72–73
heart rate, 9
    low (bradycardia), 19, 23–24, 25,
        209
Hebert, Katie, 162–64, 166–68,
        171–72
HELLP (hemolysis, elevated liver
        enzymes, low platelet count)
        syndrome, 213–14
Helm, James, 148–49
Hess, Julius, 70–71
Hess bed, 104
high blood pressure, 35
Hispanics/Latinas, 8, 246, 251
hormones, and stress, 244, 250,
        251
hospice care, 220
    African Americans and, 236
    for babies, 217, 220
    see also comfort care
Hospice de la Maternité, 57–59, 67
Hospital Corporation of America
        (HCA), 221
Hospital for Sick Children, 115
Hugo, Victor, 51, 66
human connection, 133
    bonding, 17, 91–92, 134
    see also touch
Humphreys, Jerry, 99, 107
Humphreys (Lott), Martha,
        99–102, 107–8, 111, 112, 122
Hutchinson, Alexis, 178–84, 186,
        194, 195, 202
Hutchinson, Chrissy, 178–84
Hutchinson, Isaac, 178, 179, 182
Hutchinson, Jordan, 178–84
Hutchinson, Joslyn, 178, 179

Hutchinson, Kinsley, 178
hyaline membrane disease, *see*
    respiratory distress syndrome
hyperbaric chamber, 126–28
hypotonia, 156

Ibis Reproductive Health, 266
income and socioeconomic status,
    27, 36, 246, 247, 260, 261,
    269
incubators, 39, 133–35, 151, 154
    future of, 83–96
    Giraffe, 74–82
    history of, 51–73
    invention of, 52, 93
    modern, 74–82
    as oxygen chambers, 104
    radiant warmers and, 79–80
    transport, first, 66
infanticide, 56
infections, 20–21, 24, 36
    puerperal fever, 57
intensive care unit (ICU):
    adult, 208
    neonatal, *see* NICU
    pediatric, 226, 233
in vitro fertilization (IVF), 36,
    93–94
Iowa City, Iowa, 177–78, 188
    University of Iowa, 177, 178, 180,
    182, 185, 186, 188, 190–92,
    195–202
Irish Times, 91–92
iron lung, 100, 112, 113
    miniature, 99–101, 107–11, 123
IUGR (intrauterine growth
    restriction), 3, 9, 26

Jackson-Hinds Comprehensive
    Health Center, 285
James, Arthur, 249
Janvier, Annie, 227–35, 299, 302–3

Japan, 187, 188
Johnson-Torres, Kari, 163, 168–71
Jones, Aleshia, 215–16
Jones, Jyree, 215–16
Jones, Tyree, 215–16
Journal of Medical Ethics, 92
Juntendo University, 86

Kaempf, Joseph, 200–201
kangaroo care, 16, 19, 20, 24,
    80–81, 151
Kapur, Rishi, 304
Kennedy, Arabella, 122
Kennedy, Caroline, 121, 122
Kennedy, Jacqueline, 121–28
Kennedy, John F., 122–24, 126–29
Kennedy, John F., Jr., 122, 128
Kennedy, Patrick Bouvier, 121–30,
    141
Klaus, Marshall H., 129
Klein, Jonathan, 180, 181, 195–96,
    236–38
Kropp-Kwon, Alexis, 213–16
Kuwabara, Yoshinori, 86
Kwon, Harris, 213–15
Kwon, Young, 213–16

LaGamma, Edmund, 203–6, 219
lambs, 83–85, 87–90, 93, 96, 100
Lancet, 60–61
Lantos, John D., 1, 89, 118
Laramey, April, 304
laryngomalacia, 171–72
Latinas/Hispanics, 8, 246, 251
Lazarus Case, The (Lantos), 1
Levine, Samuel, 125
Liebling, A. J., 71
life expectancy, 250
life support, 39, 45, 84
    withdrawal of, 194, 195, 203–26
Lion, Alexandre, 61–63, 66, 73
Liu, Helen Hung-Ching, 93–95

Lott, Martha Humphreys, 99–102, 107–8, 111, 112, 122
Louisiana State University Shreveport, 252
Luna Park, Coney Island, 51–52, 67–70, 71–72
lungs, 85, 101–2, 196
    alveoli in, 101, 102
    development of, 190
    eyelids and, 190
    surfactant in, 16, 101, 102, 104, 129–30, 190
    see also respiratory distress syndrome

Macbeth (Shakespeare), 53
Machine in the Nursery, The (Baker), 58, 93
Maclean's, 91
Madigan Army Medical Center, 184–85
Malawi, 160
March of Dimes, 34, 160, 258, 260, 261, 267–69, 272
Maria Fareri Children's Hospital, 203–4
Martin, Odile, 57, 58
Mascia, Kristen, 294–96
Mascia, Sam, 294–96
Mascia, Tom, 294
Maternité, La, 57–59, 67
McElroy, Steven, 89
McGill University Health Centre, 227
McLemore, Monica, 251, 273
McMaster University, 297, 299
Medicaid, 159, 161, 263, 265–68, 284
Medicare, 265
Mercurio, Mark, 201–2, 233, 234
mice, 94
midwives, 53

Miller family, 221, 222
miscarriage, 3, 5, 36
Mississippi, 260–70, 285
Mississippi Perinatal Quality Collaborative (MSPQC), 260, 261
Monaghan Company, 100
moral questions in healthcare decisions, 198–99, 233, 235–36
morbidity, 39–40
mortality, see death
mother-child bonding, 17, 91–92, 134
mother-child communication, 141–42, 145, 147
muscle tone, low (hypotonia), 156

nasal cannulas, 39, 223
National Birth Equity Collaborative, 251
National Institutes of Health (NIH), 43, 100–101, 129, 261
Native Americans, 246, 280
Nature, 83
Nazis, 65, 145
necrotizing enterocolitis (NEC), 211
Neely, Priska, 258
Negri, Gloria, 124
Neonatal Behavioral Assessment Scale, 147
Neonatal Comfort Care Program, 217
neonatology, 70–71, 72, 105, 115, 118, 173, 192
Netherlands, 187
neuroprotective care, 150
Newborn Individualized Developmental Care and Assessment Program (NIDCAP), 148, 149

New England Journal of Medicine, 187, 206
Newton, Isaac, 53
New Yorker, 71
New York Times, 107, 128
New York Times Magazine, 258
New York University (NYU), 136, 151
NICU (neonatal intensive care unit), 1, 14–15, 18–23, 26, 33, 38, 46, 47, 72, 73, 80, 90, 92, 93, 108, 111, 119, 125, 130, 134–36, 142, 151, 152, 234, 289, 293
    Als's work and, 133–34, 143, 147–51
    brain development and, 136–37, 140, 141, 150
    cost of, 55–56
    culture shift in, 210
    first, 52
    follow-up care after, 153–73
    Grove unit, 150–51
    jargon used in, 293
    Maria Fareri Children's Hospital, 203–4
    minorities and, 8
    private soundproof rooms in, 134–35
    PTSD after, 21, 292
    St. Luke's Hospital, 179–80, 183, 186
    University of Iowa, 177, 178, 180, 182, 185, 186, 188, 190–92, 195–202
    Willis-Knighton, 160–61
    see also incubators

Obama, Barack, 268
Ohmeda Medical, 79
opioid addiction, 262–63
oxygen, lack of (asphyxia), 203

oxygen treatments, 104–6, 115, 123, 124, 126, 128, 223
    visual impairment from, 105–6, 136
    see also ventilators, ventilation

Pacific Island women, 246
pain management, 135, 219–20
palliative care, 225, 227, 231
    see also comfort care
Pan-American Exposition in Buffalo, N.Y., 64, 65, 66
Papadopoulos, Chris, 113–14, 119
PAPP-A (pregnancy-associated plasma protein A), 2–4, 32, 280
Parravicini, Elvira, 216–22, 225–26
Partridge, Emily, 86
Patrick Bouvier Kennedy (Ryan), 124
Pellegrini, Cynthia, 34
Perry, Carlos, Jr., 253–54
placenta, 85, 87, 101
placental problems, 36
    insufficiency, 7, 8, 15, 31, 36
Planned Parenthood, 266
polio, 100, 113, 116
Popular Science, 94
postpartum depression, 292
post-traumatic stress disorder (PTSD), 21, 292
poverty, 160, 260, 261, 269
preeclampsia, 34–36, 263
    HELLP syndrome, 213–14
Preemie Voices (Saigal), 299–301, 303
pregnancy:
    length of, 35
    with multiples, 36, 40
Premature and Congenitally Diseased Infants (Hess), 70–71
prenatal care, group, 271–86

Prentice, Claire, 71
preterm (premature) birth,
    premature babies (preemies):
adjusted age of, 42, 154, 295
African American, 8, 34, 243–59,
    262, 263
concept of, 60
decisions about care for, see
    decisions and choices about
    care
defined, 35
disabilities from, 41–44, 91,
    155, 158, 186, 193, 207, 221,
    231–32, 234, 289–90, 297–99,
    301, 302, 305
genetic contribution to, 248–49
gestational age in, see gestational
    age
grown preemies speak for
    themselves, 297–305
hidden trauma of, 289–96
history of, 51–73
incubators for, see incubators
long-term consequences of, 33,
    289, 297–305
medically induced, 35
medical risk factors for, 35–36
in Mississippi, 260–70
monitoring and prediction of, 37
NICU and, see NICU
prevalence of, 32–33, 34, 243,
    260
profound, 38
racism and, 243–59
radiant warmers and, 79–80
spontaneous labor or rupture of
    membranes, 35
stress as risk factor for, see stress
treated differently than other
    patients, 232–33
treatments and outcomes,
    39–44, 46
two ages of, 42
ventilators for, see ventilators,
    ventilation
viability and, see viability
warming methods for, 54
weight at, 38, 40, 59, 72
working conditions as risk factor
    for, 261, 266–67, 270
progesterone shots, 36–37, 263–64,
    268–70
PTSD (post-traumatic stress
    disorder), 21, 292
puerperal fever, 57

Qatar, 79
quality of life, 235, 239, 297, 300–302

race, genetics and, 247–48, 254
racism, 8, 34, 36, 160, 279
eugenics and, 64–65
in health care, 254–55
in Mississippi, 261, 269
preterm birth and, 243–59
stress caused by, 36, 245, 250
see also African Americans
radiant warmers, 79–80
Radziwill, Anna Christina, 125
Radziwill, Lee, 125
Raffel, Dawn, The Strange Case of
    Dr. Couney, 63, 65, 68
Recht, Louise, 67, 69
Reese, Corey, 110
Reluctant Genius (Gray), 103
respiratory distress syndrome
    (RDS), 15–16, 66, 68, 70, 99,
    104–7, 115, 120, 122, 130,
    141, 190, 223
gasping for breath, 219–20
as hyaline membrane disease, 99,
    104
Kennedy and, 121–30
oxygen for, see oxygen treatments

research on, 129
surfactant and, 16, 101, 102, 104, 129–30, 190
respiratory support, respirators, 39, 136
see also ventilators
resuscitation, 39, 45, 46, 54, 177, 180, 181, 184, 185, 187, 195, 206, 221, 222
chest compressions, 180
decisions about, 189–92, 199, 201, 227, 231, 232
retinopathy of prematurity (ROP), 106
retrolental fibroplasia, 106
Retrolental Fibroplasia (Silverman), 106
Rh disease, 14, 114
Rodman, John, 54–55
Roe v. Wade, 27, 47, 197, 199
Rogers, Elizabeth, 150
Romanis, Elizabeth Chloe, 92
Rossetti Infant-Toddler Language Scale, 163
Royal Victoria Hospital, 227–28
Rushdie, Salman, 306
Ryan, Michael S., 124–28
Rysavy, Matthew, 185–87, 194

Saigal, Saroj, 297–303
St. Luke's Hospital, 179–80, 183, 186
Salinger, Pierre, 124
Scandinavia, 271
Scientific American, 64
sensory stimulation, 135, 138–40, 165
Shullman, Charlie, 210–13, 226
Shullman, Graham, 210–13, 215, 216, 226
Shullman, Laura, 210–13, 215, 216, 226

Shullman, Lilly, 210–11, 213
Shullman, Stella, 210–11, 213
sickle cell anemia, 248
SIDS (sudden infant death syndrome), 23, 155
Silverman, William A., 72, 106
Sinclair, Jack, 189–90
Sklamberg, Felice, 136, 151–52, 156
social determinants of health, 261
Society for Pediatric Research, 118
socioeconomic status, 27, 36, 246, 247, 260, 261, 269
SOLARS, 251
South Dakota Urban Indian Health, 280
Sowek, Emily, 290
Sowek, Liz Rodriguez, 290–94
Sowek, Noelle, 291, 293–94
Stahlman, Mildred, 99–101, 107–11, 118, 120, 122, 123, 125, 129
Starr, Karen, 76–77, 89
Stead Family Children's Hospital, 177, 180
steroid treatment, 6, 37, 40, 180, 190, 196, 231
stillbirth, 34
Strange Case of Dr. Couney, The (Raffel), 63, 65
stress, 36, 243–45, 261, 269, 284
group prenatal care and, 271, 272–73, 279
hormones and allostatic load in, 244, 250, 251
from racism, 36, 245, 250
weathering and, 250
stridor, 171
sudden infant death syndrome (SIDS), 23, 155
Sudduth, Amy, 161–62
Sullenberger, Chesley, 77
Sundell, Håkan, 109

SUNY Downstate Medical Center, 114
surfactant, 16, 101, 102, 104, 129–30, 190
Sweden, 187, 188
Swyer, Paul, 115, 117, 118

Tarnier, Stéphane, 56–62, 78–79
Teigen, Chrissy, 292
Texas, 221–22
therapy, follow-up, 153–73
Thompson, Wengora, 267–69
tocolytics, 37
touch, 80, 135
    kangaroo care, 16, 19, 20, 24, 80–81, 151
tracheotomy, 224–26, 236–38
treatments and outcomes, 39–44, 46
trial of therapy (trial of life), 194–95, 204, 206
Trisomy 18, 218
tubes:
    endotracheal, 120, 183, 191
    feeding, 39, 59, 60, 182–83
Tucker Edmonds, Brownsyne, 192–93, 202
Twain, Mark, 53
22-week babies, 18, 40, 45–46, 88, 90, 91, 177–202
23-week babies, 18, 40, 46, 88, 89–91, 177, 184–88, 195, 199–201, 206
24-week babies, 40–41, 45–46, 59, 80, 88, 137, 177, 232

umbilical cord, 85
UnitedHealthCare, 272
University of California, San Francisco (UCSF), 150–51, 251, 273
    Preterm Birth Initiative (PTBI), 255, 257, 258

University of Iowa, 177, 178, 180, 182, 185, 186, 188, 190–92, 195–202
University of Pennsylvania, 119, 141–42
University of Tennessee, 80
University of Tokyo, 86
uterus, 36

vacuum jacket, 103
Vanderbilt University, 99, 100, 109, 111, 272
ventilators, ventilation, 39, 122, 123, 126, 130, 141, 223, 224
    Bird, 116, 129
    CPAP, 17, 25, 39, 223–25
    high-frequency, 196
    iron lung (negative-pressure), 100, 112, 113
    iron lung, miniature, 99–101, 107–11, 123
    long-term, 238–39
    long-term, with tracheotomy, 224–26, 236–38
    positive-pressure, 39, 108–9, 112–20
Vermont, 267
viability, 38, 45–47, 60, 90, 198
    abortion law and, 27, 46–47, 90, 197–99
    eyelids and, 190
    moving line of, 190–91
    parental discretion zone and, 46
    see also gestational age
vomiting, 158

WakeMed, 148
Walsh, John W., 121–22
warming methods, 54
    radiant warmers, 79–80
    see also incubators

Warren, Elizabeth, 258
Weill Cornell Medicine, 94–95
Wexler, Irving, 114
Whitton, Gerald Brent, 134–35
Wiener, Alexander, 114, 115
Wilkinson, Dominic, 195
Williams, Hope, 255–57
Williams, Serena, 255
Willis-Knighton Health System, 134, 159–61

womb, artificial (biobag), 83–96
womb rooms, 133–34
Wonder, Stevie, 106
working conditions, 261, 266–67, 270
world's fairs, 62–64
World War II, 145

Yale-New Haven Hospital, 161
Yale University, 272

About the Author

SARAH DIGREGORIO is a freelance journalist who writes for various publications, including the *Wall Street Journal*, the *New York Times*, *Martha Stewart Living*, *Food & Wine*, and *Saveur*. Her work has been included in the *Best American Food Writing* yearly anthologies three times. She is also the author of a cookbook, *Adventures in Slow Cooking: 120 Slow Cooker Recipes for People Who Love Food*, with a foreword by chef Grant Achatz, which was published in 2018 by William Morrow.